Visions of Paradise

Bernhard Grzimek
Reinhold Messner
Leni Riefenstahl
Herbert Tichy

Visions of
aradise

Contents

Paradise today

Everyone has their own idea of paradise. Some people are lucky enough to find their particular paradise here on earth. What's more, some are able to continue savouring it, with no fear of the banishment Adam and Eve suffered from that first paradise, the Garden of Eden.

When we asked Reinhold Messner where his paradise was he answered without any hesitation, 'My paradise? The mountains.' Everyone, he says can share the experience of the mountains, regardless of how high the peak or how quickly the last few hundred feet have been climbed.

For Professor Grzimek the idea of paradise conjures up the many National Parks of Africa, huge conservation areas, still as unspoilt as the Garden of Eden.

Leni Riefenstahl found her paradise in 1956 among the Nuba people of the Sudan who allowed her to share their simple, happy way of life. Then she discovered another paradise on the trail of Jacques Cousteau in the depths of the ocean.

While still a young man, the Viennese climber and explorer, Herbert Tichy, found his paradise up under the roof of the world in the countries of the Himalayas. Lands then closed and difficult of access, such as Tibet, Nepal and Ladakh, are still breathtakingly beautiful.

Today man's remaining Gardens of Eden are in constant danger from man himself, but *Visions of Paradise* offers some rare and privileged glimpses into the beauty that survives.

Illustrations

Reinhold Messner
My Paradise – the Mountains
Most of the photographs in this selection, both colour and black and white, are by Jürgen Winkler, Wolfratshausen, and first published here. Except:
Gerhard Klammet, Ohlstadt:
    p. 10/11, 12/13 top, 14/15
Bildagentur Mauritius, Mittenwald:
    p. 12/13 bottom

Bernhard Grzimek
My Paradise – the World of Animals
Toni Angermayer, Holzkirchen:
    pp. 74, 81, 84 bottom left, 98/99
Willi Dolder, Pfäffikon/Switzerland:
    pp. 64/65, 68/69, 70 bottom right, 86/87, 88/89 top, 88 bottom left, 90/91, 92 bottom right, 94/95, 101
Bernhard Grzimek/Okapia, Frankfurt:
    pp. 57, 70 bottom left, 92 bottom left
Root/Okapia, Frankfurt:
    pp. 58/59, 70/71 top, 84/85 top, 89 bottom, 93 bottom
Okapia, Frankfurt:
    pp. 60/61, 71 bottom right, 72, 75, 79, 80, 82/83, 88 bottom right, 92/93 top, 97, 102
Hans Reinhard/Angermayer, Holzkirchen:
    pp. 62/63, 77, 78, 96, 100
Günter Ziesler/Angermayer, Holzkirchen:
    p. 85 bottom

Leni Riefenstahl
My Paradise – Africa
All the photographs in this selection, both colour and black and white, are by Leni Riefenstahl, Munich, and with few exceptions first published here. Except:
Horst Kettner/Riefenstahl, Munich:
    p. 105

Herbert Tichy
My Paradise – Himalayan Country
All the photographs in this selection, both colour and black and white, are by Jürgen Winkler, Wolfratshausen, and first published here. Except:
Herbert Tichy, Vienna:
    p. 198

Leni Riefenstahl
On the Trail of Jacques Cousteau
All the colour photographs in this contribution are by Leni Riefenstahl and first published here. Except:
Horst Kettner/Riefenstahl, Munich:
    pp. 201, 217
Ludwig Sillner, Mainburg:
    pp. 218/219, 220/221, 222/223, 224
Foto Keystone, Hamburg:
    p. 241

Front endpaper:
Jürgen Winkler, Wolfratshausen

Title page:
Willi Dolder, Pfäffikon/Switzerland

Back endpaper:
Hoffmann-Buchardi,
Anthony Verlag, Starnberg

Layout and picture editing: Art Studio, Munich
General production: Mohndruck Graphische Betriebe GmbH, Gütersloh
First published in German, © *1978 Saphir Verlags GmbH, Munich*

British Library Cataloguing in Publication Data
Grzimek, Bernhard; Messner, Reinhold; Riefenstahl, Leni and Tichy, Herbert
    Visions of Paradise
Printed in Germany.   ISBN 0-340-27220-1

My Paradise
the Mountains

Reinhold Messner

When we were children the Villnösstal in South Tirol was our world. Among tree roots and empty haylofts, with the crumbling cemetery wall alongside, we lived in an adventurous kingdom, wildly and yet harmoniously completed by the towering peaks of the Geisler chain. I grew up in Villnöss, the son of a village schoolteacher. I was only five when my father took me with him on my first mountain climb, on the Sass Rigais. I do believe today that this first mountain – the whole ascent, the excitement, the peak itself – so engraved itself on my mind that I was a climber from that moment and never lost that first enthusiasm.

I was five, too, when for the first time we were allowed to accompany our parents and help them carry things up to the *Alm*, the high summer pasture with its hut. When we got there we found a paraffin lamp had been left behind at our first resting place. 'Well,' said our parents, 'if you go back down and find the lamp, we'll let you stay a few days up here on the *Alm*.'

So we did and it was during those days that we climbed the Sass Rigais as well.

While father managed the whole business with calm deliberation, to me it was all a mystery. When I came out of the wood on to the meadow of the *Alm*, there stood the mountains, looming up directly behind it. The whole range was so huge, so gigantic. I still particularly remember that over the last few yards before the peak a strange man who was descending towards us stopped to help, and carried me part of the way. And then there was the view from the top, a thousand metres below us the Gschmagenhart *Alm*.

These mountain walls are 800 metres high, a normal height for the Alps, nothing sensational, but for me at five bigger than anything I could have imagined. They made a bigger impression on me then than did the Rupal Wall on Nanga Parbat later, the highest ice wall in the world. That I was to see when I was twenty-five.

Meanwhile, as a schoolboy, I stood under the East Face of the Little Fermeda, which is graded III for difficulty. Though I was still rather in-

experienced and deeply impressed by the exposure of the route, I felt no fright when under the second chimney overhang my father offered me the lead. I was to be first on the rope!

'Take great care and climb slowly.' He emphasised his words. 'And when you're up, hang your belay on a sound spike of rock.'

When we had finally reached the top I thought, 'Well now I really must be looking almost grown up!'

From my tenth to twelfth years I teamed up with my next younger brother, Günther, who was very important in my development because he was a skilful climber and had at least as much staying power as I had. Günther and I then did climbs on our own, for which we still had not the necessary experience or much in the way of previous briefing, and which we knew that our father hadn't yet done in his twenties and thirties. We were terrifically proud to enter against our names in the Sass Rigais summit book the North Face route by which we had come up. When our father learnt about it he was just as proud as we were, although he himself had always failed to manage it.

It was in these years that we developed that instinct for the terrain, for the rock face, and for difficulties, which was later again and again to help me out of extreme situations.

In Villnöss there was nobody but us who climbed, absolutely nobody. The farmers and neighbours had only one word for it, 'Those boys are crazy.' We were young, inexperienced, hungry for adventure. We wore wide cord breeches and washed out anoraks, and our rope was hemp. Today it seems to me we probably worked off our puberty up there in the mountains. We couldn't dance, never bought flowers for the girls, and blushed when by some chance we roped up a girl with us. After early mass we marched from home to the approaches and in the evening we marched back home. We looked unmoved at the farmers who shook their heads at our rucksacks. Besides, we never visited a café and looked down on all those who spent their Sundays there.

Marmolata, the Langkofel Range, and Sella. In the background the Central Massive of the Alps. Reinhold Messner climbed the Marmolata di Rocca by traversing, for the first time, the Plattenschuss, a steep wall leading up to the summit. The lower half of this wall is not shown in our picture.

Flowers are part of the mountain experience. In spring, alpine anemones are to be enjoyed, and in summer the much-beloved edelweiss whose flowers are found in the Swiss Valais as high as 3400 m, in the Himalayas even at 5000 m above sea level. Until late in autumn, the large heads of silver thistles decorate the mountain pastures.

As we had no transport, we simply set out early in the morning, climbed a mountain, and came down again by one of its faces. That's how we became a single rope team, two brothers, and gradually acquired all that knowledge and experience.

I was by then a student. Günther had a job in the bank at Bruneck in the Puster valley, and we used to meet at home with my parents. He came from work, and I was studying for an exam, and suddenly we said in a joking sort of way, 'Let's drive to the Sellajoch and try the North Face at the second tower.' I had often looked at it, 250 metres high and never climbed before, a black, perpendicular wall. So we drove there.

We arrived at three and soon after three we started the climb. For the first rope's length I didn't really even need a belay. While Günther was tying himself on I was already climbing. The second rope's length was the most difficult. It was so difficult that after half an hour I began to wonder. We had a go, we went up, we came down again. We were ready to call it a day. It was late, and we had only come for the fun of it. If we make it, we make it – if we don't, nothing's lost. Then I did after all manage it, the most difficult pitch. But the upper section was still to come and that was far the longer. Well, three-quarters of our route was up this single vertical wall. But suddenly, absolutely everywhere you needed it, there was a hole, a hand-hold or a foot-hold, so we were able simply to balance our way up like gymnasts, climbing into the evening sun. We completed this first ascent in a few hours. Today it is rated as one of the finest climbs in the whole of the Dolomites.

Even at that time it was a dream of mine to make a first ascent of a really difficult rock face in the Dolomites – solo, myself alone. And in 1969 I achieved it, when I was twenty-four.

I had already tried this face I had in mind once with my brother. We had climbed the lower section and then bivouacked. It's a wall over 800 metres high, on the South Face of the Marmolata di Rocca, and very very difficult. At our first attempt we had had new snow on the second day, impossible to get any further. In the end we had a real problem even to get back off this wall without having to be rescued.

And then a year later I meet this German climbing party in the Dolomites who tell me that's where they're going. *They* want to make that first ascent which was my dream.

That I could not allow – after all it was *my* idea!

So I rang up at least ten people – my brother was too busy – one after another to ask if they'd like to come, I needed a partner for this first ascent, I couldn't let the other lot beat me to it. One after another, friends I knew well, or climbers I didn't know so well who were quite famous, one and all they said, 'The wall's quite impossible!' Impossible too it was for another reason – I refused to use expansion bolts. If I had agreed to take expansion bolts some of these climbers would have been ready to join me. But it's a principle of mine in rock or mountain climbing to use artificial aids only for belaying and not for making progress, because I always say, If I can't *climb* the pitch, I'll stay down!

In despair at not finding a companion, yet wanting to do it all the more, because I was just mad keen on this one wall – I simply drove down there and tried it alone. I started climbing, once again there were no immediate problems, I reached the lower section, I knew that already, and bivouacked at the bottom of the wall.

Next day I had relatively good weather and started to climb the wall alone. Well, that's quite a different matter of course, when you've got nobody there to belay you. In the middle of the critical upper section of the wall I reached a place where I could never have believed I'd make it. For a long time I tried one place where I just couldn't get through. Then further to the left I made a left traverse and there hammered in two pitons to belay myself, and then I found a fine crack with just room for my finger-tips, and so climbed what was really an uncommonly difficult pitch.

What followed was decidedly easier, and only just under the summit I had a short passage which was difficult. So I finished the climb all in one

day, and the other climbing party didn't even try it after that.

Of course I was pleased, I had got there just in time, I'd brought it off, a first ascent, solo, at the highest grade of difficulty. When I think back to this and other climbs, there were some pitches when I can, still today, bring back the exact sensations and feelings I had when climbing. Generally too I'm a man who thinks very aesthetically about first ascents. A first ascent for me is some sort of a work of art, something of that kind.

First I study the face in photographs and pictures, then I look at the real thing and with my mind's eye draw a line across the face. I tell myself, that's the way I must go, left, right, up that crack. It may be a crooked line or a straight one. Such a line, which doesn't exist, is seen only by experienced climbers!

Take the world's greatest, the South or Rupal Face of Nanga Parbat, where I made a first ascent in 1970 with my brother Günther. It's the highest rock and ice cliff on earth. From its foot in the Tap fields it soars to a height of 4500 metres. It was a wall that nobody had climbed before us.

Climbers before had challenged the Rupal Wall, they'd made their attempts, they'd thought out the line to take. We went and found a new line, partly, too, rediscovered the old line. The line at first was just an idea. Would it go there or wouldn't it? No-body knew. But in our brains already it was a line – across a wall 4500 metres in height from foot to summit – and the summit of Nanga Parbat is 8125 metres high. And to think that we can take a mountain wall like that into our headpiece and draw a line across it there!

And what matters to me is to draw that line with elegance.

A dream first ascent from the point of view of the line, for instance, is the Heckmair Route up the North Face of the Eiger. The line is so ideally placed that for a good climber it's the only

Roof on the Aiguille du Midi. In the background the Dome du Gouter.

26

possibility, the only free possibility where you can get up without iron-mongery.

The sensation of happiness in climbing is equally great for all who love mountains. I don't think I feel more happiness or greater satisfaction than the next man, or stronger feelings.

The important thing in climbing is to have the courage to tackle an extreme situation – not a life-and-death crisis – but a situation of uncertainty. Without uncertainty you can't have a deep, strong mountain experience. It's only in the uncertainty of an extreme situation that your body and your feelings – I'm sceptical about intellect – your body and your feelings become receptive of sensations and experiences important for a human being.

My mountaineering isn't important for humanity, even though I have climbed Everest without oxygen. But for the person who does it, mountain-climbing can be a vital need.

Ability and experience have a different range. If a person has very little experience and not much ability, he soon gets into an extreme situation. If he becomes afraid too, he should turn back. Fear is a natural barrier. The extreme situation is just as much of one for him as for me under the summit of Everest, when I can hardly get through. But I am convinced that his feelings and experiences are just as intense when he is crossing a wind gap in the Dolomites which I sometimes walk over in gym shoes in course of training.

When I look back on my feelings as a ten-year-old when I started up the East Face of the Little Fermeda – which now I could do in a Sunday afternoon in walking shoes – which I don't do however! – my excitement was the same then as I feel now for Nanga Parbat, which I should now like to climb solo. In those days I had very little experience, well, you might say none at all, and so a tiny little mountain and quite a small face were enough to excite me and agitate my nerves, or at any rate to plunge me into a domain of experience which is still important to me and which I still hanker after today. Only today I must look for relatively big mountains to reproduce the same excitement.

27

*On the Col de Peuterey,
Mont Blanc.*

In the Dolomites an extreme situation for me was undoubtedly an experience on the Langkofel in 1969, when I did the North Face, immediately above the Grödner Valley, solo by a very difficult route, the Soldà.

First I visited it three times and twice I wouldn't risk it. I had a look at the face and went back home, because I couldn't bring myself to start the climb. The third time I went there about noon. I always have more courage in the afternoon than in the morning. I know that now, and I knew it then.

Well, I reach the end of the main difficulties. And what do I see? Crumbling rock, very small holds, it's

clear I'll never make that! That last pitch is much too hazardous. I get out a piton to hammer into a crack, and begin to hammer, so as to loop the rope through, so as to belay myself.

I'd barely driven the piton in two centimetres, when the handle of the hammer broke! I still had the hammer in my hand, the shaft was broken and the hammer still hanging to the wood, but I could hammer no more pitons with the head alone, the metal lump, and so I had no way left to belay myself. I did some quick thinking. I can't possibly wait to be rescued. If I start shouting, shouting for help, it'll be night at soonest before anyone comes. Yes, and I'll never be able to hold out

that long, not a hope!

So it was make or break, I had to do the pitch. I don't know if I'd have done it otherwise. At first I was afraid – but the moment I left the stance, when I told myself, Well, I've got to do it, or fall off, end up eight hundred metres below . . .

It was the lesser evil, I had no alternative and just stormed through, into life.

In retrospect it proved to me that when people get into a really tough extreme situation, they can surpass themselves.

And it's true, I once found myself on a pitch that was even more difficult – I remember it as the most difficult I

ever climbed. I was climbing with my brother, and in that case we did have a belay. It was in the Dolomites, too, on the Heiligkreuzkofel.

This bit of face was a compact, vertical, partly overhanging pillar. We went left up three ropes' lengths, that's 120 metres, then we found a traverse, a spur in the wall which I'd spotted with binoculars from below. Here we went forty metres to the right, almost horizontally across the vertical-to-overhanging face. We climbed a bit further, and then it came, the crux. Above a great overhang I ascended a bit vertically with artificial help – that is, a few pitons I drove into holes – I only accept pitons driven into natural

holes, nothing else – then I climbed free up a crack as far as a small spur four to five centimetres broad. I could actually stand and rest a bit by stretching my toes.

If you stand for long on tip-toe, after half an hour you can't stand at all, the calf muscles are under such strain that you tremble, and if you tremble you lose your stance. So there I was standing on this spur, I was just able to get upright, I was able to do it from below, it was a question of equilibrium. There was hardly anything to hold on to, I just groped my way upright against the wall without holds. But I did have a stance, I could stand, and so I gradually raised myself upright without 29

falling backwards, although it was vertical. Above me was a sort of overhanging bulge, only about six metres high, and I saw above that was a crack. Broad enough, I could see, for me to ascend with, if I could reach it. But there were the four to five metres in between, they were smooth, they looked as smooth as the wall of a house.

I got the message, it's no go, I can't climb that. I tried, but I felt immediately, it's throwing me over backwards. If I try with only one foot to climb just ten centimetres higher than the spur, I shall fall off the wall backwards.

When I realised it was no go I called down to my brother, 'I'm coming down again, we'll have to abandon, it's impossible!' Then in my attempts to get down I suddenly realised that was no good either, I'd fall off backwards, because I had no holds, and there was no way to climb back down off that spur again. When I realised I was about to fall, I thought to myself, 'What shall I do?'

Just to jump, simply to jump down, I hadn't the courage. For me it was an important discovery. When you see no way out, you still don't, as matter of concrete choice, risk the jump.

Then for half an hour or so I was trying, first to climb upwards, just to get a start, and finding it was no go. While I was trying I thought to myself, 'Perhaps you'd better go down', and when I tried downward I thought to myself again, 'Perhaps upward would be better.'

30   *In a snow-storm on the Gran Paradiso.*

A Himalayan giant, the 6856 metre Ama Dablam, one among the highest range of mountains in the world with its fourteen eight-thousanders.

Neither seemed possible. And because both were impossible, because for me there was no question of jumping, because I'm human and full of fear, I simply tried to climb upwards, I said to my brother 'Watch out, this time I'm really risking everything!' And by these small roughnesses, which as it turned out were wider than hair-cracks, where one could hold on almost by one's finger-nails, I hauled myself up. I risked it and suddenly I was on top and I don't know to this day how I got up there, suddenly there I was on the upper part of the pitch where I was able to make progress, to my great relief, as you may imagine.

Page 33
German climber on Manaslu at about 6000 metres.

Page 34
Few Himalayan expeditions are without their Sherpas, the porters who carry the equipment and provisions for the high-altitude camps. The porter in this glacier crevice is carrying a thirty kilogram load barefoot at a height of 5500 metres.

Page 35
Ice towers with annual layers in the Himalayas make it possible to read off the exact measure of the precipitation for decades.

The German-sponsored Manaslu Expedition of 1977. Camp II: After heavy hauling, these sherpas enjoy a good rest in their tent, 5500 m above sea level.

The interesting thing is that nobody has ever climbed this pitch after me, though it's exactly ten years ago this year. Nobody's done it yet.

All that kind of compensating act on the summit of Everest, crying yourself back into equilibrium, so to speak, where the weeping is only the outward expression, that's simply the emotional reaction – there's a lot going on there in our feelings.

And I'm convinced these outbursts are only possible on the summit. Ten metres below the summit, and the effort would have been perhaps just as great, the tension beforehand would have been the same, but we shouldn't have had these outbursts of emotion. Up on top there the intellect was absent, it had ceased to function, and so the emotional outbursts could happen. And I suppose the pressure of these outbursts reaches such a pitch when one has had some such dream beforehand or some such idea of having to go on to the end.

When I was down below again in the Everest Base Camp, it lasted a whole day. Both of us, Peter Habeler and I, we were really 'high', you could almost say. But then I suddenly felt something missing, not that I was really unhappy. If you look at it intellectually, it's easily understood. I had this huge idea, important for me, which had pursued me for years. Everest without oxygen! It had been all along a sort of utopia – will it be possible or won't it? Hoping and doubting.

Now suddenly this utopia was gone, I missed it, there was a hole there. In the Everest Base Camp I received permission for Nanga Parbat, the first time I'd received permission to climb an eight-thousander solo. At an earlier time this dream had been more important in my thoughts than Everest without oxygen, only then Everest took first place because I had the actual permit to climb it, and Nanga Parbat was pushed into the background. Now that I had the permission and felt this hole, I at once took in the next idea to fill it.

The conquest of loneliness, that's the main human problem. In 1975 Peter Habeler and I climbed an eight-thousander, Hidden Peak, without high-altitude porters or a rope. We didn't tie on to one another, each climbed for himself just as if he'd been alone. But a partner was there, one you could see, like looking at yourself in a mirror, a neighbour opposite. Even though one had no conversation with him, for eight hours at a stretch, he was just there, and that was enough.

To do without this *one* human being is after all much more difficult than doing without all artificial aids.

Why do people climb mountains? There are many answers to this question. I have the feeling, quite generally – and that goes not only for me – that the mountain experience is made up of three great strands of experience.

First is the experience of achievement which is a primal need of every human being. I have never met anyone who didn't seek some sense of achievement. Even among the Danis, a New Guinea people, I have watched the boys at archery. There a boy of fifteen can hit the trunk of a banana tree at a distance of thirty metres and put an arrow right through it. Out of five shots he scored four hits. Then I tried and missed by ten metres. The young Danis were delighted that we were so unskilful with bow and arrow. These boys couldn't read or write, but they did have a sense of achievement. It was their answer to the question, who is the best shot?

Many young people today reject achievement altogether. True, western man in our industrial society is often forced into achievements which are no longer natural. But for that reason simply to reject *all* achievement and to say, that's not natural, that's not normal – I'd never do that. I would say, the sense of achievement is a central experience for every human being.

But achievement is only one strand of experience in mountain climbing. A second is the romantic experience, nature that is, which speaks out of the macrocosm and microcosm like the sun's rising and setting. That's a whole heap of things. It can also be something very small, a blade of grass outlined against the flying cirrus clouds on the horizon.

But the most important strand of experience, at least for me, and every 49

climber will be differently possessed by one or other, the most important for me is the visionary experience. This visionary experience is the very thing which occurs in a critical situation, where we suddenly acquire knowledge which cannot be got from reading books. It is knowledge from experience.

Fundamental experience, fundamental knowledge can only be gained from life, and certain definite kinds of knowledge can only be gained from getting into certain definite situations. It's not a question of luck or ill-luck but of whether you dare to put yourself in such a situation.

These visionary experiences can go so far that somebody for a short space of time – a minute or a second – understands the cross-connections of the entire world, grasping with simple intuition and naivety that it's all logical and a matter of course whether one lives and why one lives.

Every human being, at least every thinking human being, every reflective human being sooner or later comes to doubt the sense of his or her existence. But for all that I'm against the idea that it's the mountains which impart to me the sense of my existence and that that is why I go climbing. I only know that while climbing I have had moments, now and then, when I have intuitively understood, what sense *my* life has.

But I couldn't describe that today, and say, 'My life has sense, because –' But there are, in fact, momentary flashes where I simply know what sense my life has or what value my life has in the world, how big the world is or how big Everest is. That's a thing you can only do in these extreme situations. And that's what I call the visionary experience. The most important experience.

The spirit of morning in the Dolomites. Left Monte Pelmo, right the Civetta. 51

I'm absolutely convinced the mountain experience doesn't depend on the difficulty or the height of the climbs but at most on the capacity of the individual person to have sensations at all. We're all of us made differently, one is more achievement-minded, another more of a romantic. So this one, accordingly, looks more for the romantic experience, while the first is more for achievement. I do think, however, that every human being has three main experiences. What I think important is that everyone has that possibility of experience when climbing.

So that is the purest form of the man-mountain confrontation. Once upon a time it was the mountain which played the principal part, then the mountain was conquered. Man had the adventure, the difficulties were the price you paid for standing on top – and today I say it's the other way round.

The mountain's conquered, well and good. It's completely unimportant whether the mountain's conquered or not. It makes no difference to the human race whether you stand on top of Everest or not. The only thing to be explored is the human being on the way, up or down, by him or herself. Whether the mountain's unknown or not, whether it's a white patch on the map, makes no difference – let it remain so.

For myself, what interest me are the white patches in the human being. Only it isn't a case of saying when I'm climbing high, I discover things up there which are important for humanity. They're only important for me.

I don't know how I'll behave, I myself am excited to know. I go to it with great casualness, I tell myself, if I bring it off, it's fine, if I don't, I don't. I go climbing today almost as I went climbing when a boy. When a boy, I simply saw slopes, those steep rock faces up there, and we just climbed to the top without ever having heard a word about Alpinism.

A person is simply put into the world, or is suddenly there, has a definite life-span to dispose of, a life of his or her own to play with, that's how I see it. The more conscious we are that we're going to die, that our lives are limited – most people know they're going to die some day, but they're not quite conscious of it – the more intensively we can live. And we cannot rightly play with our lives until we are conscious that they are limited, that we shall die sooner or later. Then it becomes unimportant when we die, quite unimportant – but that's a long way from meaning that I play with death.

By playing with life I mean to play with its possibilities, to play with its thoughts and dreams.

If I were a man who had no more dreams I should feel very sorry for myself.

My Solo Attempt on Nanga Parbat

It was on June 30th, 1978 that I was seen off at Munich Airport. On August 7th, without any artificial aids, neither oxygen cylinders nor pitons nor ropes, I set out from the main camp. I hadn't yet made up my mind whether I should climb Nanga Parbat from the south or the west. Finally I decided on the West Flank.

On the day of my climb, about five o'clock in the morning, there was a sudden earthquake which so shook the side of the mountain that part of it broke off. Luckily I had the route well memorised, I knew of another possibility after the partial collapse of my climbing slope.

On the night of August 8th–9th I bivouacked at 6400 metres.

August 9th. I did not reach the summit of Nanga Parbat till four o'clock in the afternoon, after I had to fight my way for part of the climb up to the waist in snow. Up on top as 'summit book' I anchored a metal capsule containing the first page of the Gutenberg Bible with my name and other evidence that I had in fact been up there. Also I reeled off one complete spool of black-and-white snaps from the summit.

On August 10th I was descending by a wall almost 4000 metres high and got caught in a storm. I waited a whole day and then, on August 11th, climbed down the fall line of the peak. During the climb I lost a piece of my thumb through frostbite.

Nanga Parbat was for me the fulfilment of my last great Alpine dream, to conquer an eight-thousander solo, without superfluous artificial aids. The sporting achievement for me took second place, I wanted above all to find out how a person behaves in situations of extreme risk.

Summit Needle of the Salviccia in the Urner Alps.

Reinhold Messner
was born in 1944 in the Villnöss Valley in the South Tirol. He is head of an Alpine School he founded in St. Magdalena and his books translated into English are: *The Seventh Grade: most extreme climbing* (1974); *The Challenge* (1977); *The Big Walls* (1978); *Everest: expedition to the ultimate* (1979); and *Solo Nanga Parbat* (1980).

One look at his list of climbs is enough to tell us why he is acknowledged to be the best climber in the world today.

1970 First ascent of the Rupal Flank of Nanga Parbat (8125 metres).
1971 Expeditions to Iran, Nepal, New Guinea, Pakistan, East Africa, and several new climbs in the Dolomites and in the Carstensz Group (New Guinea).
1972 Manaslu, South Face (8156 metres), first ascent; trip to Noshaq (7492 metres) by the Hindu Kush.
1973 Further first ascents in the Dolomites (Pelmo, North-West Face; Marmolata, West Pillar; Furchetta, West Face); trip to Nanga Parbat area.
1974 Aconcagua, South Face (6959 metres), first ascent; forced to turn back on Makalu (8481 metres); alone with the Hunzas in Pakistan; Eiger, North Face in ten hours; Matterhorn, North Face.
1975 Expedition to Lhotse, South Face (8511 metres) in the Himalayas. Hidden Peak, North Face (8068 metres) in the Karakoram in a rope-party essentially without artificial aids.
1976 Mount McKinley, first ascent of the Midnight Sun Face; Ortler, West Pillar, first ascent; trip to Annapurna.
1977 Expedition to Dhaulagiri, South Face.
1978 Expedition to East Africa, Breach Wall on Kilimanjaro, first ascent. First ascent without oxygen of Mount Everest (8848 metres) on May 8th, with Peter Habeler; first ascent of Nanga Parbat (8125 metres) by the Diamir Face on August 9th, solo, without oxygen or technical aids.
1980 Climbs Everest solo.

Translated from the German by Christopher Holme

My Paradise
the World of Animals

Bernhard Grzimek

Bernhard Grzimek

Many children take an intense interest in animals, but later, as they get bigger, most of them lose it; I am one of those who have never lost it. I began, like so many children, with rabbits and guinea-pigs, and progressed to a number of goats. As a boy I bred special strains of bantam hens; I sent them to shows and won some awards.

The only thing my mother would not hear of was a cat. When, in spite of that, we were given a little kitten, I laid it beside her every day when she took her afternoon nap, by her head on the sofa, until she became accustomed to that too.

When my brother brought home a snake, she said either the snake went, or *she* did. Dejectedly, my brother, Notker, let the snake go again; next day, however, there was still one in the terrarium. Notker said defensively: 'You did say I should get rid of *the* snake – but there were two . . .'

At that time my great longing was for a pig. My mother wouldn't let me have one; so I pestered her from morning to night. When I even whimpered outside the lavatory door, 'Mummy, a pig!' it was too much for my half-sister, Barbara, who was twenty years older than me and just on a visit. She gave me a smack and told me off, at which I retorted, 'Silly goose, you can't say anything to me, you're only my sister!'

In my youth almost every family had relatives in the country, to whom one went in the holidays. There one could milk cows, help with the harvest, play in the barn when it was wet, roast potatoes out in the fields, drive and ride horses, turn the hay, collect eggs – something new every day.

We bathed almost daily in the stream behind the village, in which one could also catch crayfish. At the village blacksmith's one could tread the bellows and watch for hours how the horseshoe was bent, red-hot, and then placed, still hot, on the trimmed hoof, which stank abominably of burnt horn.

Holidays in the country – they were a true paradise for us children. We had our own pony-trap, and also a boat on the pond, in which there were also a lot of crayfish. There I learnt how to catch frogs, by making a bit of coloured material dance about in front of their heads with the help of a switch and line. They jump at it and bite hard for a moment on the supposed butterfly – at that very moment one has to give them a quick jerk out of the water on to the land. Once there one can catch them by hand. Naturally we let them go again.

My family and my teachers put up with my passion for animals with great tolerance. I joined the local association of small-animal breeders and began to breed pedigree fowl, especially bearded bantam Antwerps. For this I constructed in a timber yard, with permission from the woman who owned it, five pens with five little chicken-houses. I sent my fowls to Oppeln to a poultry exhibition and won a prize – a whole coffee service with pictures of bantams. Though it was not at all in good taste, I was at least as proud of it as I was forty years later of winning an Oscar for *The Serengeti Shall Not Die*.

For the nature-study classes, I often took along small animals, sometimes at the request of the teacher, frogs, blind-worms, mice, salamanders. Once it was a hedgehog. My classmates thereupon gave me the nickname 'Hedgehog', which I have kept, throughout my schooldays, right down to the present time – probably because it is very much easier to pronounce [*Igel* in German] than the difficult name Grzimek. So the hedgehog became my heraldic animal.

Whenever I was allowed to travel with my mother to Breslau, I pestered her continually until we went to the zoo. For me the main attractions of the Silesian capital were the zoological gardens and – the swimming bath.

At home in Neisse at that time, I had a whole lot of rabbits. Every day I had to cut grass for them from the roadside and also make hay. I collected horse-droppings from the street in a sack for my little garden. In addition I rented part of the old garrison cemetery from

Page 57
Professor Bernhard Grzimek with elephants and young rhinoceros in the Serengeti.

Page 58/59
A picture as in the Garden of Eden: Burchell zebra grazing under acacias in the bush. In the background, Kilimanjaro.

Page 60/61
Part of the magic of the animal world in Africa is the breadth of the landscape and the moods of the time of day and year. Giraffes taking their evening stroll.

Page 62/63
A family idyll: lioness with her cubs.

Amongst big game species, the rhino, too, is endangered by human shortsightedness and greed.

Cow antelopes roam the open steppe. These animals are very gregarious and form large herds.

This impala (or blackheel antelope) grows its first horns. Cape buffaloes have earned a reputation as the most ferocious animals of Africa. More humans are killed by this type of buffalo than by lions.

the leaseholder and there I grazed my two white hornless Saanen goats.

In this way I could contribute a little to the feeding of our family, for at that time we were after all fairly poor, because all wealth had disappeared, of course, through inflation after the First World War.

Every Wednesday in Neisse there was a market on the Ring. Not only vegetables, eggs, butter, cheese and fruit, but also hens, ducks, geese, dogs, goats and pigeons were sold at open stands. Some time before the main break I used to ask to leave the classroom, but instead of going to the lavatories, I would go to the weekly market, and see what there was to buy there or what I could sell of my own produce. The money for it I sometimes borrowed from the old vegetable-woman, Berta, with whom I also left my animals, before going back again, like a good boy, to my secondary modern school. The teachers never noticed this extra-curricular activity.

My greatest desire was to be able to sit on horses. That, for me, a small-town boy from an impoverished family, was something almost unattainably remote. But, in a very round-about way, I managed it all the same.

One of the members of our association of small-animal breeders was the proprietor of a little printing works, where I learnt to set type. I could pick the lead types from the letter-cases and assemble them into the right words like a regular compositor – even blindfold.

I used to draw up texts for this man, for his customers' posters and advertisements, and I wrote pamphlets for him in secret which he then distributed. Among these was one on the famous Therese of Konnersreuth, a girl who had wounds like Jesus Christ on hands and feet which bled regularly and who was at the time creating a tremendous sensation throughout Germany.

In this way I built up a credit balance with the printer. In return for this credit balance I produced posters for a former soldier who had opened a riding school. So I learnt to ride free of charge, which at that time only the very rich could afford.

Riding did not become a popular sport in Germany, of course, until long after the Second World War, when horses had been replaced as working animals by cars and tractors. At that time I was not allowed to ride through the suburbs, which were inhabited mainly by workers and *petit bourgeois*, for to them someone on horseback was provocative, recalling the *junker* squirearchy. Angry remarks would be made and stones might even be thrown. The motor-car of an industrialist, on the other hand, which was a lot more expensive, upset nobody.

After the rudiments of riding had been instilled into me with military precision in the former army riding-hall, I was allowed to ride outside as well – usually alone. The proprietor had forbidden me to gallop too much and bring the horse back wet with sweat, because that cost too much in fodder. So I usually raced madly at the start of my ride out, at an extended gallop across country, over hedge and ditch, and then for the last three-quarters of an hour ambled back in very leisurely fashion. As a result, the sweat and lather dried up again; in addition I brushed the horse down at the end and polished its coat smooth with a cloth, so that my gelding returned to its stall bone-dry.

This gelding, incidentally, could be difficult; if one spurred it too hard, it stood up steeply on its hind legs and fell over, backwards and to one side, so that one couldn't avoid coming off. Admittedly, he only did it once with me.

As a child I wanted at first to be a coachman, then a Franciscan friar, and later an officer; finally I settled for farming. At that time I was always concerned with agricultural, domestic animals only, hardly at all with wild creatures. I always thought it would be the same throughout my life. I was very fond of animals, rather in the way of a farmer. Naturally I wanted to take up a profession that had something to do with animals. So it occurred to me to study veterinary medicine, in order to become a veterinary surgeon. It was only a few decades since veterinary training had been elevated to a full university course. There were at the

The praying mantis, so called because its front legs, with which it catches its victim, are raised as if in prayer.

The gorilla, despite its ferocious appearance, usually has a gentle disposition.

time very few veterinary faculties at the university. The nearest to my home town of Neisse was at Leipzig.

But what was to become of my little black friends, my bearded bantam Antwerps? They had to go with me, just as they had gone with me earlier in their home-made travelling chicken-coops to relatives in the country during the long holidays.

From Neisse I put an advertisement in a Leipzig paper, through which I found an allotment-garden with a nice summer-house in Leipzig-Mockau. The owner was unemployed – it was in the spring of 1928.

So I duly studied veterinary medicine and zoology, became a veterinary surgeon and later the director of a zoo, and have remained with animals continually to the present day; I have not regretted it for one moment in my life.

After taking over as Director of the Zoological Gardens in Frankfurt in 1945, I always had the feeling that one cannot look after animals and house them correctly if one has no idea of how they actually live in their own habitat, in freedom.

After 1945, however, it was very difficult at first for a German even to get out of Germany and into the former German colonies, which had, of course, since the First World War, belonged to the victors. It was a complicated business, but we finally managed it. These experiences in Africa were not just scientific work, with new findings and unforgettable encounters with wild animals in their old freedom, such as I have described in detail in my books.

On my journeys in Africa I was brought together with peoples who were still living in a very primitive way – that too has unfortunately changed in the last twenty years. One could thus share the simple, genuine, village life with the black population, live with them, get to know their daily round, take part in their celebrations.

Once on the Ivory Coast we had the great good fortune to be invited by the chief of the Bauls to visit his village and the other villages of his tribe. In this way we got to know the fine old life of tribe and family, which has since disappeared so quickly. No-one can convince me that Africans have been made happier by technology and Europeanisation.

The Bauls at that time tilled their fields in common. A committee of village elders decided how much each was to receive from the harvest for his family. In the old Africa there were no orphans. If both parents did in fact die, the nearest relatives would have lost face if they had not at once taken over the children and brought them up in the same way as their own.

At the palaver, time was taken to discuss a dispute thoroughly and arrive at a sensible decision. In those days one could go into any hut and share the meal.

Because of my interest in these people and their life, it was not long before I also came into contact with the black politicians of these countries, got to know their problems and fears, and could sometimes do more for the development and safeguarding of the great national parks in Africa than official European institutions.

My work for the great national parks in Africa can only be rightly understood in the light of my attitude to animals, which is somewhat different from what is often assumed. Even as a child I realised that a good many effusive sayings, such as, 'Anyone who is fond of animals must be a good human being', are not true.

I see myself to some extent as a part of nature, see, too, the dilemmas in nature and also the horrors, know them and accept them for what they are.

The animals and we human beings are the parts of one nature. We have the same physical components, we humans eat practically the same food, breathe the same air and are in fact very much more a part of nature than most people realise. At university I often used to say jokingly to doctors that in a way they were specialised vets . . .

Because we are a part of nature, the idea that man should rule over animals, make them subject to his will, is very dangerous. Through his most recent developments, through industrialisation at breakneck speed, man is becoming much more remote from animals and, what is far worse, in destroying the animal kingdom as a

White-backed vultures. As eaters of carrion they belong to the 'sanitary service' of the bush.

The short muzzle and large 'expressive' eyes give the three-toed sloth an almost human appearance. It is known also as an ai, from its typical, two-syllabled mating-call.

Page 81
Of all monkeys, the face of the mandrill displays the most gorgeous colouring.

Page 82/83
A herd of African elephants. Unfortunately these creatures tend to eat even the bark of trees, which then die.

Page 86/87
The green mamba. Mambas are the most poisonous snakes of Africa, but luckily they are very timid.

White herons like to stand on hippos to have a good view around.

Hippos grow enormous tusks with which they are able to inflict terrible injuries to enemies as well as each other.

Crocodiles usually take a rest with the mouth open to cool their breathing air.

Cape buffaloes enjoy their sand-bath.

The shabrack-backed jackal feeds on cadaver and small prey. Jackals are closely akin to wolves and dogs with whom they cross-breed.

The male lion loves to play pasha, leaving hunting to his womenfolk.

African wild dogs set against the sinking sun. Also known as hyena dogs, they do their hunting in packs.

Flamingoes sift their food out of shallow waters: algae and other minute forms of life.

They breed in colonies. To protect their young against floods, flamingoes build their nests on top of characteristic mounds formed out of lime.

Grey plumage is typical of the young rose pelican.

The crowned crane is one of the most impressively plumed African birds.

Springboks live in the dry, open veldt north-west of the Cape of Good Hope. When the animal jumps in flight, the black stripe on the flank twitches vividly to frighten off pursuing predators.

Page 90/91
Cheetah or hunting leopard. With a speed of about 115 kph over short stretches it is the fastest land animal.

Page 94/95
The long-tailed or green monkey. These graceful creatures much prefer swinging through the branches of trees

Page 96
Typical of the Massai giraffe are its spots in the shape of vine-leaves. In the background, Kilimanjaro.

whole, and with that nature as a whole, we are in the last analysis destroying ourselves. This is something which up to now has only been realised by far too few people.

I see animals and human beings as parts of one great whole, and this great whole in nature, which man, it is to be hoped, will not finally destroy. That is the paradise in which by associating with all animals, quite regardless of whether they are domestic animals, animals in a zoo, or animals in the nature reserves of Europe or distant parts of the globe, we can share.

When one keeps thousands of animals captive in a zoo one often racks one's brains about how their old freedom can best be replaced.

But how do these animals live when they are free?

For that one must first observe them in their habitat. The zoo directors of the nineteenth century never went to Africa or other overseas countries. This may have been due in part to lack of money, but even more to the fact that a zoo director simply couldn't leave his work alone for years on end.

When one travels for weeks by

steamer across the sea, and then for weeks in a small boat up African rivers, when one has to go on safari with porters, months and years go by. Today anyone can fly to Africa in a few hours.

In addition, animal psychology and the study of animal behaviour were then still in their infancy. Hardly anything had been written on the observation of wild animals living free.

Ever more and more I felt an urge to see the brothers of my zoo-inmates in the free state.

To our parents and grandparents animals were as self-evident as the four walls of their houses. They lived with them, horses drew their carriages, waited on the streets, pigeons pecked in their backyards, the girl collected fresh eggs from their own chicken-houses. Now animals have become rare. Our children get to know them now only in films, at the zoo and perhaps during the holidays. What has been lost gives food for thought, and its value is recognised only later.

Today especially I should like to emphasise that my life's work has been and remains fundamentally the conservation of nature, in which wildlife protection can only be a part.

The conservation of nature seeks above all to prevent individual species of animal on our globe from being exterminated, to prevent tracts of land from being ravaged, laid waste and its life broken up, as has happened already through stupidity on a large scale all over the world.

The conservationist believes that the fate of nature depends on man as the only superior creature. Conservationists, hence above all scientifically active biologists, are in general made of somewhat sterner stuff than pure wildlife protectionists.

A biologist knows that the struggle for existence in nature at large is hard and that wild animals by no means live in a 'paradise'.

I see nature as it is. Out of the two million large animals that now live in the Serengeti again – in the last twenty years they have increased five-fold under our protection – at least a third of the antelope calves die every year, since they are torn by crocodiles, killed by predators and in part sometimes starve, so that they drown by the hundred in crossing rivers and lakes.

That too is nature. If the predators were to be shot, the other animals would multiply on a vast scale, destroy the whole soil and its growth, and then, because of their excessive numbers, starve and die out.

So, in the struggle for life among wild animals, gruesome things undoubtedly occur everywhere, though no unnecessary or conscious torments. As a matter of course, the conservationist, in the same way as the wildlife protectionist, sets his face against cruel forms of hunting and treacherous animal-traps, which are already forbidden by law in most countries. In practice the two functions are often quite inseparable, for instance in the protection of fur-bearing animals. The cruel hunting of the pretty baby seals, the Greenland seals in the Gulf of St. Lawrence in Canada, may be mentioned as just one example. Let us

Migrating white-bearded gnus in a part of the bush ravaged by fire.

98

never forget, after all, that the only one who needs an ocelot coat is an ocelot.

Pure, unperverted nature, a bit of 'paradise', still exists today in the great national parks, for which I have fought all my life and will continue to fight. A national park is not fenced in; it is a piece of countryside in which only indigenous animals and plants live, the animals naturally without being hunted, the woodland quite without forest management, hence a piece of genuine wild land.

The many national parks, above all in Africa, are a great cultural achievement of the young states there. These peoples are proud of their heritage. For the ancient African religions had already created regions in which the killing of animals was forbidden. Thus, in the event of over-hunting and extermination, the gaps could be filled up again by the immigration of wild animals from the protected regions.

The African man has mysterious bonds with the animal soul. His soul changes over from time to time into animal bodies. After death it often lives on in animals. These notions are probably one of the reasons why the new national parks in black Africa have a far better basis than in the countries of the whites.

Poor African countries sacrifice several times more for the conservation of nature and the national parks than the USA or we in Europe.

And the black people who, as curators and gamekeepers, defend such a paradise, which, it is to be hoped, may one day be enjoyed by a peaceful mankind and visitors of every colour of skin, deserve our respect and appreciation.

The klipdas or rock-rabbit, which is really a coney. Despite their resemblance to the rabbit, coneys are the nearest living relatives of the elephant.

We *must* defend the national parks in Africa. In them alone will a small proportion of the continent's beautiful large animals survive the modern development of Africa. Whether it will succeed is still a question. But we Europeans cannot unload responsibility for it on to the Africans. We must help them so that the animal world of the most beautiful and colourful continent shall not be destroyed within a lifetime.

Under black rule much has been done in Africa. Tanzania, for instance, today has ten vast national parks. It has enlarged the Serengeti very considerably, and even evacuated whole villages to do so, something that was never proposed under European colonial rule, let alone done: to be willing to evacuate people to make room for animals. But the gamekeepers, the vehicles that soon fall apart in the bush, the little light aircraft for surveillance, which they pilot themselves – all this costs money.

Let us protect these paradises, which are our paradises too, by giving money and by making our politicians understand that we are interested in the preservation of nature and the animals in it.

What can association with animals give us today? It can give the many people who have been made lonely in modern industrial society personal contact with another living creature. Our grandmothers no longer tell the little ones fairy tales, but sit, carefully separated from them, alone in an old people's home or a well-equipped two-room apartment, where a dog or a cat has to compensate them for the love of their grandchildren.

An animal can compensate many young people for what our ancestors up to fifty years ago all still had. They associated not only with human beings but also with other living creatures on earth, in the way that human beings have done daily for three to four million years.

Since time immemorial it has been part of human existence on earth to live in nature alongside animals and with animals and among animals. So we must associate with animals in order to remain human beings.

Swallows are widespread throughout Africa. Their nests often consist of a pile of grass and leaves.

101

Bernhard Grzimek

was born on April 24th, 1909, in Neisse in Silesia, the son of a legal counsellor. At Leipzig university he studied veterinary medicine and zoology. He worked his way through college.

After practising for a short time as a veterinary surgeon in Berlin, he waged a successful campaign, as an inspector in the Ministry of Food, against chronic fowl infestation and bovine tuberculosis.

His scientific work was concerned in particular with the anthropoid apes, wolves and the domestic horse. He took part in the Second World War as a veterinary officer.

In the spring of 1945 he was appointed Director of the Frankfurt Zoological Gardens, which had been completely destroyed, and built them up again; today they have more than three million visitors a year.

Grzimek then went on exploratory journeys to West, Central and East Africa, Japan, Canada, Russia, Australia and South America. In 1956, with his son, Michael, he made a full-length film about animals and the primeval forest, *Kein Platz für wilde Tiere* ('No Room for Wild Animals'). Father and son carried out by aeroplane counts of wildlife stocks in East Africa and observed their migrations. While doing so Michael crashed and was killed in the Serengeti bush in 1959. His documentary film sequences, under the title *The Serengeti Shall Not Die*, was the first German film to receive an Oscar in Hollywood.

Professor Grzimek's German television series *Ein Platz für Tiere* ('A Place for Animals') has already brought in donations of more than ten million DM for nature conservation abroad and in Germany. The government of Tanzania appointed him curator of their national parks.

At the end of 1969 the government of Federal Germany appointed him honorary special commissioner for the conservation of nature and the countryside. As early as 1973, however, he resigned and declared that the goals of a conservationist could not be carried through within the federal government.

Even after his retirement in 1974, Grzimek has continued his research and publicity work. He is fighting with great vigour for the protection of the environment, in which the conservation of nature and animals is only a part.

His publications translated into English include, *Among Animals of Africa* (1970), *Twenty Animals, One Man* (1974) and his Encyclopedias of ethology, ecology, evolution and animal life.

Translated from the German by George Unwin

My Paradise Africa

Leni Riefenstahl

It was very rarely that an Arab lorry came as far as the Nuba mountains. When it stopped by the huts with its load of cotton or salt, the whole village gathered around it, everyone wanting to talk to the driver.

Although more than fifteen years have passed since then, I still can't forget a little episode I witnessed at the arrival of such a lorry. The scene is the key to one of the last paradises of my life.

One of the Arab lorry drivers gave me two flat loaves of bread: a gift from heaven, since I had had to live on nothing but *durra-brei*, the traditional Nuba food, for weeks. I had no money, I lived all alone with this native tribe in the province of Kordofan in Sudan.

I kept one of the loaves for myself. The other one I gave to a Nuba man called Natu, whom I knew well and who might be even more hungry than I was.

Natu took the loaf without haste, quite calmly, as if he wasn't even hungry, let alone starving. A couple of children were standing around us. Natu looked at them, I didn't realise why. Then he counted the children and began to break the bread, carefully and deliberately. He gave each child a piece of bread and broke off one piece for a mother and another one for the baby she was carrying. Finally only one small piece remained for Natu himself: it was the smallest piece of all.

I was moved by the fairness and the humanity of this scene. It was a symbolical expression of everything this paradise constituted: not the exotic quality, not the enchanting landscape, but the people. They made the paradise. What is it they say? Hell is ourselves . . . With the Nuba people I learnt the opposite.

Although the Nuba live in great poverty, in extremely primitive conditions – judged by the standards of our so called civilisation – I have always longed to go back there. The deprivations of their existence in the mountains don't count. Because the people are of a kind we no longer come across.

They lived in complete harmony with nature. And they lived with great social equality. Want, greed, envy or stress did not exist with them. The Nubas appeared to me to be truly happy people. They had not yet been ruined by money. They helped their fellow man without asking themselves what benefits it would bring them. They were grateful for any sympathy they got and returned it ten times over. One felt secure with their love and affection.

If a man was unable to look after his fields due to illness or while away working in Khartoum, the others did it for him as a matter of course. When I sprained my ankle, they looked after me. They built a hut so that I could live with them, they dug a deep hole in the ground – it was supposed to be a well – so that I should have water, although there was none. They shared their *durra-brei* with me.

They were certainly people with faults, too, but these were harmless and lovable. Perhaps they were a bit gossipy, still it all seemed basically innocent. There was no nasty, malicious aggression, no egotism, no crass materialism, no stress. I laughed more with the Nubas than I've done in all my life. In those days I seriously considered staying with the Nubas forever. We even discussed my future burial. Everyone wanted to sacrifice his best cow for my funeral. If I hadn't had my mother still living in Germany, I might well have decided not to go back. But every two years I returned to my paradise with the Nubas.

Anyone who has been to Africa will never lose a yearning for this continent. After Hemingway had spent his first night in a tent, he wrote in his diary: 'When I woke up in the middle of the night I just lay there listening, already longing to go back to Africa again.'

I have felt the same ever since I went to East Africa in 1956 to make a documentary film about the modern slave trade. The project failed due to a bad motor accident I had in North Kenya.

On the last day of my stay in Africa in 1956, I happened to come across an old issue of the illustrated magazine *Stern*. I leafed through it without paying much attention, when I was sud-

Pages 106/107
Shilluks on their way to a celebration. The shields are of dried crocodile skin.

Page 108 top
A Shilluk asleep. The stand under his head protects the elaborate wig.

Page 108 bottom
Shilluks with magnificent ornate wigs.

Page 109
A Shilluk. As a token of their tribe, Shilluks have a kind of 'string of pearls' tattooed over eyes.

Pages 110/111
The break-up of a Masai settlement. As soon as the shepherds of a community can't find any more sufficient grazing, the Masais wander on. While the men are already off with the cattle, on the lookout for new pastures, the women organise the removal. Among the possessions being removed are calabashes, eating utensils and wood. Wood is expensive and is brought along in bundles.

Page 112 top
Unlike the Nubas, the Masai people are warriors by nature.

Page 112 bottom
Masai men plaiting each other's hair. Strings of wool are plaited into the hair to change its structure.

116

denly electrified by a picture. George Rodger had photographed two Nuba men: an athletic ring fighter sitting on the shoulders of a friend. This was one of those moments which can change someone's life. I wanted to visit these Nuba men, I wanted to get to know these people.

Six years later I had my chance. I could join a small team of German scientists who were planning an expedition through the Sudan.

Our little group travelled in the most primitive conditions possible. I had neither a tent nor any other camping equipment. While the men always made their night camp on the roof of the Unimog car, I slept on a narrow camp bed in the open air. My full equipment consisted of a sleeping bag, a blanket, a mosquito net, a drinking mug and plate, a tiny saucepan with a spirit stove, a thermos flask and a torch. Two wash bowls was all we had between us.

I had brought something else, which the men considered to be totally unnecessary but was indispensable to me. I didn't really want to return from the Tropics with dried-out skin, so I had filled some plastic bottles with oil and tonic lotion and also brought a greasy face cream, tissues and even hair curlers. In two other metal trunks, my camera, the flashlight and a lot of photographic material were packed.

Our food was mainly rice, noodles and tea. We had no fruit or vegetables. In this area they didn't even grow bananas. Meat was scarce, since these dry areas had little game. After only a few weeks I looked rather thin. But I didn't worry about my weight. And I didn't worry about the modest diet or the long days of being jolted about over unbroken ground. And I didn't worry about the heat which reached 40° to 50°C in the shade. I was simply happy – from the first minute to the last.

I thought of a passage which I had written down after five months with an expedition in Greenland, after filming Dr. Arnold Franck's film *SOS Iceberg*. 'What is the great wonder,' I had asked myself then, 'which enthralls us so here?' And I had made my own reply: 'We're suddenly seeing different things, feeling different things. Europe's questions and secrets fade away – they don't exist any more. What provoked us at home is almost unfathomable here. Other measures, other values – and so a huge ballast of superfluous, unproductive things, which never bring any happiness, fall into the sea. Time – and with it, our whole life – is given back to us once more.'

This reply also applies to Africa.

We had to search for a long time before we found the still 'untouched' Nubas. A great deal of this tribe, even in those days, had been included in civilisation. Only in the most remote valleys of the Nuba mountains were there still Nubas who had retained their ancient customs and habits. We made our camp there, in the shade of a tree with a leafy canopy thirty metres across, and in the immediate vicinity of a Nuba settlement.

While the scientists concentrated on their work, I tried to establish contact with the Nubas and learn their language. *Joka-i* was the keyword. *Joka-i* meant: 'What is that?'

A Nuba child taught me that. With this word I could now inquire about a lot of things. If I pointed at my hair, the children shouted *nagga*. If I pointed at a finger, they shouted *maui*. Soon I could count to one hundred and had learnt the words for 'today', 'tomorrow' and 'the day after tomorrow'. In this way we soon made friends and while I was watching them I could take pictures without any trouble. The Nubas were totally without inhibitions in front of the camera. I have never ever taken such natural photographs of any people.

They were soon interested in everything to do with me, mostly, of course, my white skin, which they liked to touch, my blonde hair and, naturally, my clothes.

What I found particularly striking about the Nubas was their happiness and their great love of music. The men, boys and girls often played instruments, not unlike simple guitars. Although they were very poor, I have never known such joyful people. They are like big children; they take every chance to have a good laugh. At the same time their life is really full of deprivation: their main food is the

Nuba huts are built of stones and clay. While the walls are about three metres high, the highest point of the roof is up to five metres. All around the walls there are little holes the size of a stone 122 which give constant ventilation.

The Mesakin Nuba tribe has tattoos made with sharp thorns, fine metal knives or crystal splinters. The girls are tattooed on the breasts, upper arms and back when they have their first period. The elaborate back tattooes often have the shape of an arrow. With each pregnancy they have a new body tattoo.

durra, something in between millet and maize. They also have cattle, but not all families have a cow. The herds, which are tended by young shepherds, are very small – unlike those of other African tribes. Due to the drought, the cows yield only up to one litre of milk daily. This milk goes to the men only. Women and children live basically on *durra-brei*.

The men have this privilege because they supply the ring fighters. The ring fight is after all the focal point of Nuba life. The Nuba ring fights have a strong connection with cult and religion and are a unique ritual.

The first time I witnessed such a Nuba ring fight, I thought I had been removed to a different planet. The participants looked completely un-

real, mainly due to the fact that they rub themselves strongly with white ashes beforehand. The effect is rather like stone statues. Moreover they adorn themselves with feathers, beads, and large calabashes and paint their bodies in strange patterns. The supple movements of the ring fight thus becomes a ballet of modern sculpture and body art.

One day, much too early, I had to say good-bye, because the members of the expedition team had to return to Malakal, a small Sudanese town on the Nile. From then on I was alone, left to my own devices.

There for the first time I saw the interesting Shilluk and Dinka tribes, which belong to the Nile people. In Malakal I made the acquaintance of a

124

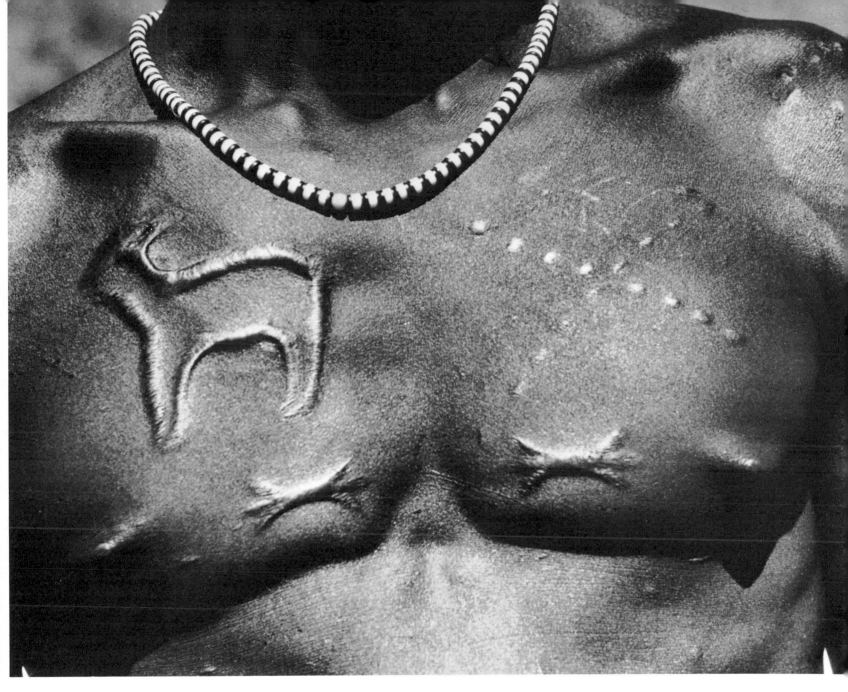

German and an Englishman who were on their way to the Congo with a Volkswagen bus. They let themselves be persuaded to take me to the Shilluks. Their capital, Kodok, is some 400 miles north-west of Malakal.

Here I got to know the famous Shilluk king, who has 107 wives and must die when he is past the climax of his vitality. The king, called Sun-God by the Shilluks, is seen as the incarnation of their god, Nyakang. If the king becomes weak or ill, they believe that this will reflect on the whole people. Their harvests will dry up, their cattle will be sick and their wives sterile. When the time comes for the king's wives to announce to the Council of Elders that their ruler is showing signs of weakness, a great celebration is ar-ranged, where the king is given alcohol and then poison.

One evening I saw against the setting sun how tens of thousands of Shilluks danced their war dances. As with the Nubas, their western neighbours, the death cult plays an important role. And this dance which I was watching was in the honour of a great dead man. The Shilluks carried heavy shields, the size of a man, made of crocodile skin, together with three spears. The air trembled with excitement, the dust gleamed a metre high against the glowing red sky. Wild drumbeat and screams were heard.

The Shilluk men who belong to the warrior caste wear elaborate wigs, made of animal hair and plaited into their own hair so closely that they can't

With the Mesakin Nuba men, tattoos are seen as tokens of courage. The more beautiful ornamental scars a young man has, the greater his chances with the girls.

The tattoos aren't always successful. In order to obtain more or less plastic shapes, they spread ashes in the wound, which can, however, also cause scars that are too thick and ugly. 125

be removed. For this reason they sleep with their heads on little stands to protect the magnificence of their hairstyle.

Apart from this, it is easy to recognise a Shilluk, because everyone has a very striking tattoo in the face to mark his tribe. It looks like a string of pearls on the forehead. This tattoo is made with a fine metal point when they are only three or four years old.

Unlike the Nubas, the Shilluks are rich, living as they do on the fertile banks of the river Nile with its good grazing land. Various anthropologists are incidentally of the opinion that they originally were related to the Watusi and Masai tribes – cousins, as it were, of these other two Nile people.

Anyhow – the Masai tribe, which I also visited in 1963, appeared to me to be the complete opposite of the Shilluk and Nuba people. In those days they were proud, unapproach-

able and full of contempt to all strangers, black and white alike. Only a few kilometres from a modern metropolis like Nairobi, they still maintained their ancient customs. Our 'achievements' did not impress them. This probably has its basis in their religion, whose convictions strangely enough are very close to those of the Old Testament.

One little episode may illustrate their inner independence and their refusal to adjust. The Masais hate to be photographed. However, I was determined to take some pictures of a Masai with a very striking helmet hairstyle shaped with red clay. He looked like an old Roman, at the same time haughty and composed. I tried to photograph him, but he put his hand up to ward me off. I still took a few shots. To my surprise, he laughed. I managed to utter the few words I knew of Masai, in an attempt to make con-

A Mesakin Nuba ring fighter (left) and a South-East Nuba with an arm knife fighting.

A Nuba ring fighter at a death festivity. Only the totally blameless are selected to kill the sacrificial cows.

126

versation. I was even more surprised when he replied in perfect English. I asked him where he had gone to school, and he said: 'In London.'

'And then what did you do?' I asked him.

'I qualified as a teacher.'

'And now you're back in the bush!'

He laughed and said: 'I like to be a Masai.'

But let us return to the Shilluks, these remote 'cousins' on the Nile.

One evening I sit together with the Shilluks round the fire. It is full moon, and I feel spellbound by the night light, as I have often been, since my childhood. I am overcome by a strong desire to return to the Nuba people. But how? Several hundred kilometres of unbroken ground are between them and me. Would my two companions undertake such a trip for my sake? The VW bus is probably badly equipped for it. But they agree, and on the second day we set out.

From Malakal we go by ferry across the Nile, but only a few kilometres further on the car sinks into the marshy ground. It's impossible to get the car out and darkness is drawing close. We're surrounded by flat steppe land. Not a hut, not a soul to be seen. The two men are at a loss what to do and furious with me for persuading them to take this trip.

I'm feeling guilty and decide to walk alone the ten kilometres back to the Nile to get help in Malakal. The sun has already set and the twilight soon follows. Although I'm in a situation which naturally should fill me with fear, I suddenly experience a strange feeling while I'm walking along, of freedom and happiness.

I hear the sounds of wild animals, the humming of mosquitoes and other insects. The air is as filled with music never heard before. The sky with its glittering stars seems like an immensely wide backdrop.

In the direction of Malakal the sky is coloured red. A steppe fire must be raging over there. That's lucky for me – the light of the fire will direct me in the darkness of the night. I see a faint light approaching me. Is it a hallucination? I stop, and the silhouette of a Shilluk riding a tricycle emerges in the darkness, I speak to him, and it turns out that he works for an American mission and speaks English. He takes me to the bank of the river Nile, looks for a Shilluk with a boat and finds one. After promising my saviour that I shall visit him at the mission, I get into the boat.

The sky, red from the fire, is now quite near. I'm sitting in the narrow boat, trying to balance, since any jerky movement of my body would make it capsize and provide a feast for the Nile crocodiles.

We reach Malakal and I mobilise some help. The hospitality in the inner Sudan is unbelievable. The following morning three military vehicles with a dozen Sudanese soldiers go back with me. The VW bus is pulled out of the marsh, and we can continue our journey to the Nubas. On the way we visit the mission, as I had promised, and spend the night there.

When we finally arrived in the Nuba region, it was late at night. We found the tree under which our camp had been before. The full moon had already appeared over the hills, but not a Nuba as far as the eye could see. My companions went to sleep. While I was still busy unpacking, I heard the first voices. Suddenly, as if they had grown out of the ground, they stood before me – I could hear them shout, 'Leni, Leni, Leni *giratzo*.' That means, 'Leni is back.' And then they approached, a few to begin with, but soon they were in their hundreds, pressing my hands, embracing me, the children plucking at my clothes, the joy indescribable. It was exactly as I had wished the reunion to be.

Sleep was no more to be considered. We went up among the hills, and in the moonlit night, a great dance festivity took place under an ancient tree. The greetings continued. I naturally had to take part in the simple dances, which went on until the moon disappeared. Then the Nubas took me back again to my camp.

Page 129
A South-East Nuba on his way to the fighting ground. The fights with the striking-ring are very tough. In order to frighten the opponent, this Nuba has painted himself with black paint.

Page 130/131
The oiled and painted young Nuba men gather for a nightly dance.

Page 132
The girls of the South-East Nuba tribe rub their bodies with oil and paint. Their fighting requisite is a whip.

Page 133
In the last of the sunlight the girls dance their Nyertum – *the love dance of the Nubas from Kao-Nyaro-Fungor.*

Page 134
The South-East Nubas have a great talent for painting their faces and bodies. Their 'masks' vary from one day to another. During the ritual dances, the men sit at the edge of a Rakoba with their eyes fixed on the ground.

Page 135
Indefatigable, the young Nuba girls dance into the night.

Page 136/137
Slightly shy but still very curious, the Nuba children watch everything we do in the camp.

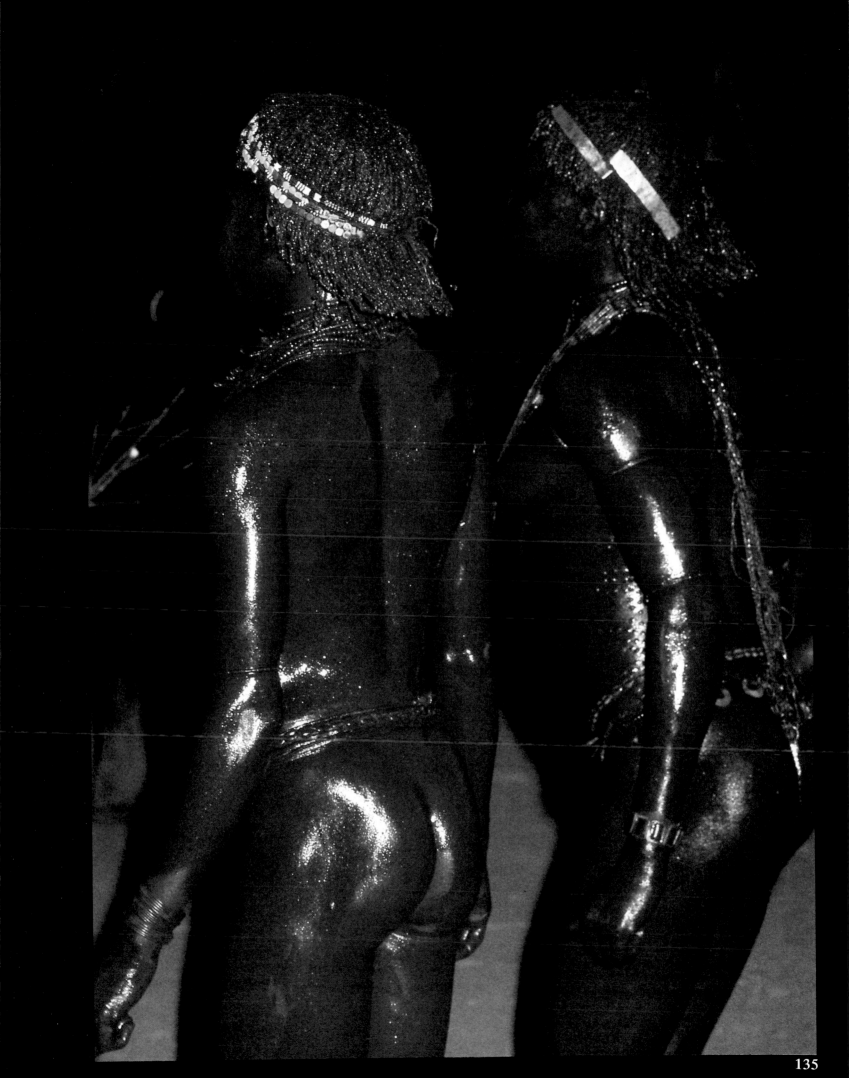

A Nuba growing tobacco. The ground is stone hard and can only be worked with iron hoes. The Nubas have to carry water from far off. Tobacco is the prime exchange goods for jewellery, spear-heads and knives. They can't forge iron themselves.

Page 138
Nolli, a Nuba child.

Page 139
A Nuba mother with her baby. The white crosses on her body indicate that she is having her period. No one must touch her during these days, since the women are then considered unclean.

Page 140/141
The initiation of a young Nuba man, an important ritual in the life of the Nubas. The bunch of coloured ribbons wrapped around the upper body of the man is the festive clothing which from now on he will wear in his fights against the strong ring fighters.

Page 142/143
A ring fight in the Nuba mountains. Thousands of Nubas gather from a distance of fifty kilometres. The women carry water and marissa *(beer) in large jugs. Arabs in white watch the spectacle.*

Page 144
Before each ring fight, the Nubas rub their bodies with ashes – an ancient ritual. They believe that the white ashes give them strength. At the same time, the ashes protect them from insects and keep the skin cool.

In the days that followed, my contact with the Nubas became even closer. But unfortunately my two companions did not stick to their promise of spending four weeks there with me. They were bored. What should I do? I paid farewell visits to the various huts, although I had a feeling I would be back again.

The farewell was even more sorrowful than the first had been. For a week I travelled with my companions through the Sudanese bush landscape to Wau, the capital of the province of Bahr el Bazal, where I said good-bye to the two men.

In the Wau area I took some pictures of the Dinka people. I met an Arab merchant who let me travel north with him again. Another merchant let me travel with him for another 200 kilometres. On the fourth day I was back at my old place with the Nubas in Tadoro.

During this third stay the great moment, which I had long anticipated, finally arrived. A messenger from a remote Nuba commune invited us to a death festivity. Glowing heat beamed from the sky as we made our way there. I felt dried up and for the first time didn't hold out. Heat and thirst had made me totally exhausted. The Nubas made a long pause but I didn't recover. When they realised this, two strongly built women put me in one of their large baskets which they carried on their heads and marched on. I was too exhausted to wonder about it.

A Nuba burial ground. The Nubas bury their dead in a central spot near the village. On the graves they place large clay calabashes.

147

At last we stopped. The women gave me a massage and put damp cloths over me. Soon I felt better again. We spent the night in a Nuba village.

The following morning I woke up feeling worn out. When I stepped out of the hut, I saw thousands of Nubas who had gathered in the beaming sunlight. They carried spears, shields and bright flags. To me it looked like a painting of a mediaeval army camp.

Thirty-six cows were to be sacrificed for the dead woman, an unusually large sacrifice, considering the poverty of the Nubas. Three large Nuba men, adorned with pink feathers, had been selected to kill the cows by piercing their hearts with their spears. After the animals had been killed, the various ceremonies began. Later on the meat was distributed between the distant relations.

Now the ring fights started, for all their cruelty never brutal, but always supple and full of grace like a dance. I took as many photographs as I could. I didn't get any rest, because the real death celebration soon started.

The procession with the dead woman, who was carried into the valley, came closer. At first came a Nuba with a white flag. The Nubas carrying the stretcher with the dead woman suddenly began to walk in zigzag. This was to ward off evil spirits. Then the dead woman was carried to her brother's hut. Only her left hand showed under the shroud. Everyone sprinkled some ashes on it. They brought gifts, large pots of *durra* grain and *durra* beer and also strings of beads, tools and bark cloth.

When I came out into the open again, I could see that the Nubas had begun to dance. They had painted white figures on their black skin – white stars are symbols of grief. The dead woman was carried out of the hut. A relation had already entered the burial chamber that had been dug out. The dead body was handed to him, and then the gifts were placed in the grave.

Once more the ashes were sprinkled over the body and after that the opening of the grave was covered with a large boulder. Earth was heaped over it. It was now in the middle of the night. Around the grave scattered mourning relations were crouching, making their profound wailing and whining noise. They remained there for hours, and daybreak slowly set in . . .

A last paradise . . . two years later I was back with the Nubas. I received a message that my mother in Munich was seriously ill and I left everything to go back there. But when I arrived in Munich six weeks later, my mother had already been buried . . . that was one of the worst blows ever dealt to me by fate.

I returned to the Nubas deeply shaken. Fifty kilometres away from 'our' village they came to meet me, asking me how my mother was. I told them she was dead. Then they embraced me and wept with me. In the village everyone mourned.

The two Germans with whom I had travelled to Africa never even asked about my mother. That is the whole difference. The paradise – that is the people. Or not at all.

Leni Riefenstahl

was born on August 22nd, 1902 in Berlin, the daughter of a merchant. After her matriculation examination she took classes in painting and drawing at the Art Academy and subsequently trained as a dancer with, amongst others, the expressionist dancer Mary Wigmann.

The first mountain film by Dr. Franck, *The Mountain of Fate*, evoked her interest in film and in natural experiences of mountains. She learnt skiing and climbing and became well-known as the leading actress in *The Holy Mountain* in 1926. Then followed *The White Hell at Piz Palü* (1929). In 1930, she had great success with *Storm over Mont Blanc* and the skiing film *The White Inebriation*, in which Hannes Schneider, the famous skier, was her leading man.

The first great success of her own was *The Blue Light* in 1932. She was script writer, producer, director and leading actress. It was about a nature child whose pure love of the blue light of the rock crystal is ruined by the greed and vulgarity of the inhabitants.

In 1933, Leni Riefenstahl took part in an expedition to Greenland organised by learned scientists. One result was the film *SOS Iceberg*, made with the assistance of the later Air Force General, Ernst Udet, and some famous Alpinists.

In 1934 she was commissioned to make the film *The Triumph of Will* on the Nuremberg National Party Day of the NSDAP, and in 1936 she made a film on the Olympic Games in Berlin for the International Olympic Committee. There were two parts of it, the *Celebration of Beauty* in 1936 and the *Celebration of the Peoples* in 1938. For this film she was subsequently awarded the Olympic gold medal by the IOC.

Leni Riefenstahl travelled to Africa for the first time in 1956. She then concentrated her work on Africa and kept returning to the Nuba tribe in Southern Sudan. Today she speaks their language. Her photographic reports of this tribe first appeared in the world's leading magazines. When the first picture-books of the Nubas appeared in 1973, the eminent photo-

graphs were admired all over the world.

In the meantime the Olympic films were shown abroad with a success that not even Leni Riefenstahl had expected. In 1956 her Olympic film was classified as one of the ten best films in the world.

In 1972 she learnt to dive, first in the Indian Ocean and then in the Red Sea, developing her own style of submarine photography. The complete list of her medals and awards won between 1932 and 1975 is too long to account for here. Her major books translated into English are: *The People of Kau* (1976); *The Last of the Nuba* (1976); *Coral Gardens* (1978) and *The Films of Leni Riefenstahl* (1978).

Translated from the German by Ann Henning

150

My Paradise
the Land of the Himalayas

Herbert Tichy

When I think of Nepal, it isn't the golden roofs of Kathmandu, or the view, from that mountain fringe, of the gentle hills of Tibet – though it's true even they are 5000 metres high – but it is Muktinath which comes to my mind.

Muktinath is a small place of pilgrimage north of Annapurna, and when today's tourists are lucky they can photograph there the three holy flames which are probably what gives the place its religious importance.

In 1935 I was working in Kashmir at my geological doctorate. With the coming of summer the pleasures of mountaineering took precedence over book work. The sacred cave of Amarnath, nearly 4000 metres up among the beautiful mountains of Kashmir, and dedicated to Shiva, the god of destruction, was the goal.

With my friend Chatter Kapur, who later accompanied me to Tibet, I set out. The pilgrims from the heat of India make formidable preparations, they're not accustomed to altitude, cold, and snow. All we did was to stuff our rucksacks with sleeping bags, climbing irons, and some provisions. We passed an ice-cold glacier lake in which a thousand-headed snake, well-disposed towards the pilgrims, has its home. In gratitude you're supposed to wash in the waters of the lake or even bathe in them. Copiously we did so. When we at last reached the cave the pilgrims had disappeared, but so had our provisions.

The interior of the cave was filled with ice. Water dripped from the ceiling and formed standing pinnacles of ice recalling the *lingam*, the Hindu phallus. When our eyes had become used to the half-darkness, we recognised in the background a sadhu, a holy man who, despite the cold, was wearing very little. We humbly touched his feet and had a friendly conversation with him.

His needs seemed to be well supplied, for the pilgrims had left him an abundance of nourishing offerings, which we eyed greedily. The holy man was amused and said it was a topsy-turvy world. Instead of our contributing to his subsistence, it was he who must feed us, as he proceeded abundantly and good-naturedly to do so.

On leavetaking he pressed on us juicy apples and hard chupattis, and gave us his blessing. That was my first encounter with a Himalayan holy man.

In Srinagar College I occupied a tiny room finished with little more than a chair, table, and wooden bed. It was not usual and also unnecessary to lock the doors. Some weeks after our visit to the cave, when I returned from an extended geological excursion, I found a sadhu in my room. He reminded me of the hermit of Amarnath. He had the same long white hair and a spiritual face. He was sitting cross-legged in the Buddha posture and meditating. After some hours his spirit returned and he said he would be living with me for a while. He shook his head with a smile when I indicated the poverty of my lodging. He did not need luxury.

My guest must have been an important man. When he preached in the open air, usually under a shady tree, he was surrounded by a respectful crowd. My room, too, had more visitors now in one day than previously in a week, and we received more offerings than even the ravenous hunger of a student could cope with. The police and tonga-wallahs (cabmen) who till then had not thought much of me, because I rode a motor-cycle, began to greet me and give me friendly smiles. I never discovered why the holy man so distinguished me with his good will. Perhaps my unpilgrimlike behaviour in Amarnath had been spoken around among his 'brethren' and amused him? Perhaps he thought I might become a useful pupil?

At all events he invited me to accompany him on a pilgrimage to holy Muktinath in Nepal. At that time, 1935, Nepal was a closed country. I was on fire with eagerness and quite willing to sacrifice two or more of my terms of study. We should be going the whole way on foot and living on alms.

What an opportunity of getting to know the life of a holy man! I, however, would need the Maharaja's permission. But that was no obstacle, said the holy man, he was a friend of the Maharaja, who would not refuse him this request. We wrote to Kathmandu and the answer came back short and sharp. 'His Highness has found it inconvenient to give you permission.'

That put an end to my dream of Nepal, and the holy man moved on in the same matter-of-fact way in which he had entered my life. He comforted me with the words, 'You, too, will come to Muktinath. I know you will.' I had not yet learnt to put my trust in the words of a holy man, and thought them no more than a kindly pat on the back. But scarcely two decades later I really was in Muktinath. In 1953 I had got permission from the Nepalese government to cross Western Nepal from Kathmandu to the Indian frontier. I could choose my own route and climb any peak I liked. All I must promise was not to cross the Tibetan frontier.

I could have taken a European companion, but preferred to travel alone with four Sherpas. We travelled for four months and covered some thousand kilometres. During the whole time we had no news of the great world outside or even of Kathmandu and did not see a single European. I was myself the first European to travel part of the route, but astonishment was always held within bounds and I was never molested from curiosity. In Kathmandu, today thronged by thousands of tourists, there were then only a handful of Europeans – the Swiss geologist, Toni Hagen, who knows Nepal better than anyone, the revered American, Father Moran, and of course the indestructible Boris Lissanevich, who was just opening a hotel and without whom Kathmandu is unthinkable. In those days if one met another European in the street one stopped, without offence, to ask them all about themselves, where they were from and where they were going.

North of Dhaulagiri we discovered the Barbung Valley, the Barbung Khola. I thought I had never seen anything more overwhelming – in the south the sky-cleaving peaks of Dhaulagiri and Churen Himal; slopes on which the grass turned to autumn gold, and trees whose leaves turned bright red; every day a radiant blue sky; inhabitants whom we believed even when they told tales of flying lamas.

In print I have called the Barbung Khola 'The Most Beautiful Valley in the World'. Today I often meet Alpinists who talk of their expeditions starting from the Barbung Khola. Every so often one of them claps me benevolently on the shoulder, 'You know, you didn't exaggerate all that much when you called it the world's most beautiful valley.'

In those days we were living on the country and carried a few tins and Nescafé only for the peaks. Apart from that, however, there were some items of equipment we considered vital – medicines, bandages, warm clothing for high peaks, crampons, ice picks, climbing rope – and the expedition cash box, which had already given us a lot of trouble.

I had been warned. In the interior the people would accept no paper money, only coins. We were a modest party, but five men for four months would require a sum which in coin would amount to a load of thirty kilograms. On occasion a local financial genius would try to put us in a panic – in the next valley our Nepalese coins would no longer be accepted, there only Indian were in use, but he, the genius, was ready to oblige us by changing our money without a qualm. Sooner or later, I would probably have been taken in by some such trick, but Pasang's peasant canniness preserved me. 'Wait and see,' he said, 'at the worst we can come back to the financier. A few days more or less in a four months' trip don't matter.' Pasang's instinct proved correct, right into Western Nepal, where they did turn out to prefer Indian money. But we had been warned of that by the Governor of Jumla, the last 'town' of any size, and he was not a broker but a gentleman.

The proverbial hospitality of the Nepalese was frequently a problem. At first I encouraged it because I wanted to learn as much as possible of 169

land and people at first hand. The Sherpas, too, preferred a convivial evening with villagers to a lonely encampment, and they used their cosmopolitan charm to procure for us a cucumber, a mug of milk, or other delicacies. Usually I did not sleep in the interior of the house where it was stuffy and smoky, to say nothing of the lice and bedbugs, but on the flat roof. The roof could be reached only by a 'climbing tree'. This was a tree-trunk in which notches had been cut, at what I must say were very distant intervals. The local people climb up and down them with the agility of monkeys, without even using their hands. As for me, I had difficulties. They were terrifyingly high (up to five metres) and had a tendency to revolve on their own axis. Also the notches were tiny and trodden out. Even the Sherpas did not have an easy time with these trees. The hospitality of the house-owners was not usually confined to milk and potatoes but included huge quantities of home-brewed beer or spirits. After a lively evening the climb up to the roof was undoubtedly one of the most dangerous moments of our journey. No doubt men the worse for drink in the Himalayas have their own guardian angel . . .

We had some strange meetings. North of Annapurna we met a Tibetan Lama who was 'taking a stroll' from Lhasa to Kathmandu. All he had with him were a tiny Pekinese, which angrily snapped at us, and a small travelling bag to carry the dog in when the way was marshy or led through a brook. He was equipped as we should be in Europe when we take the dog for a walk. We admired his light-hearted style of travel in an area where we needed a small expedition.

What difficulties could there be, he asked, with the people everywhere so kind? As a Lama, it is true, he was sure

The village of Ghandrung in Nepal is 2000 metres above sea level. The different forms of building are determined by the climatic differences and the availability of wood or stone as a material.

of a welcome, but we, too, found the people uniformly obliging. He had hardly any advantage over us. A little subdued and thoughtful, we continued our journey.

It was altogether a thoughtful journey, if only because of the *chorts*. These are massive Buddhist cult buildings, somewhat recalling our chapels, which are often erected at the heads of passes. From their walls the 'eyes of the Buddha' gaze out upon the land as a symbol of all-knowingness and all-goodness. This all-knowingness haunted me, because the eyes, often only suggested by circles, follow you for hours, questioning, mocking, reproachful, according to your own state of mind. Often it was two or three hours before we reached the head of a pass. All this time you are eye to eye with the all-knowing one and your own conscience. What way is your own life taking? How many stupidities have you already committed? Will you do any better in future? Long conversations between the eyes and yourself, while you gasp for breath and only wish you had reached the pass. Not just a religious range of mountains, these Himalayas, but a thought-provoking one!

In one small hamlet on the outer walls of the houses we found artistic, harmoniously coloured spirals which would have done credit to a young Picasso. They had been done by the villagers, not for ornament, so the people explained, but to keep off the evil spirits. In the next village, half a day's march distant, they laughed at the superstition of their neighbours, they had no use for spirals but relied on a carved wooden tutelary god who menaced us from the top of a wall. In this region of North-Western Nepal people cling to their old nature religion. Hinduism and Buddhism have not yet reached these valleys.

Marpha in the Valley of the Kali Gandaki. Between the valley floor and adjacent Dhaulagiri the difference of altitude is 5800 metres. The strong wind which blows daily imposes this cramped style of building, so that one can walk through the village from roof to roof.

With the Sherpas I spoke a mixture of Hindustani, Nepalese, Sherpani, and English, not much use for philosophic discussion but quite satisfactory for everyday use. With the local people I conversed with the same mixture but leaving out the English words. So I was extremely surprised when a peasant one day greeted me with a friendly, 'Good morning, Captain Sahib.' It was a Gurkha. The Gurkhas are not ethnologically speaking a tribe, but named after the small town of Gurkha from which Prithvi Narayan conquered the Vale of Nepal. Today *all* Nepalese soldiers are called by the name Gurkha. As British mercenaries in both world wars (50,000 in the first, 250,000 in the second) they were in Cyprus, El Alamein, Tobruk, Monte Cassino. Their bravery is known all over the world.

After the first disappointment that I was not a Captain Sahib, my new Gurkha friend beamed at hearing that I was a German (I did not expect him to have heard of Austria). Yes, he had known many Germans, in North Africa, where he had fought with the British Army against them. Then they had sailed over the sea and come to lands with mountains, *chang* ('wine'), and big cows. This was the order in which he named the things which had impressed him.

My informant received a pension of £50 a year, quite a big sum for Nepal. He had seen half the world, killed white men, and spoke serviceable English, perhaps he had marched in the victory parade in London. Now he was back in his little village that had never yet acquired the use of the wheel. Why had he brought no revolutionary ideas back with him? Only memories, a few decorations, and a pension! A mere mercenary who had come home or a philosopher? He proudly showed me his miserable fields and his thin cattle. For the milk he gave us he would not accept money.

I remember our last camp before crossing the frontier into India. We had accomplished something we had not been all that certain could be carried out. We were correspondingly happy and also wistful. The adventure of the new was over. Tomorrow we should be in India, back among streets and railways. Didn't I have the feeling that I was losing a paradise?

We had put up our tents on a stubble field close below a wretched farmhouse. The owner, who had generously given us permission, squatted like a slightly amused pasha on the terraced field above us and watched our preparations. He was surrounded by his family, tiny to half-grown children, a bloated wife, and a skinny old crone, all unwashed and clothed in rags. I had got to love the country and pity seized me. The moment I was back in Europe, I would try to initiate some measures of aid, some form of sponsored development for the peasants whose kindness had never failed, all those four months. While I was having these thoughts, the peasant was talking excitedly to Pasang.

'What did he say?' I asked Pasang.

Pasang had a fit of coughing. From long experience I knew this was a sign of acute embarrassment. I had to press him very hard before he would come out with it: 'He's so sorry for you, that you have to sleep here on his field and have no house and no goat and no cows of your own. He is considering what he can possibly give you as a present.'

The differences in how we see things!

On our tramp through Nepal we hardly ever saw a real holy man or hermit, only occasionally some pilgrims to the holy places around Kathmandu. Solitary walkers prefer the Garhwal Himalaya, especially the valleys there around the sources of the Ganges. Near Badrinath I was once in company with a holy man for quite a long time, and he seemed to live on 'air and God'. I never saw him eating anything and the people told me he had spent the winter in Badrinath, totally deserted and buried in snow as it was, without any provisions. At night – the temperatures used to fall just below freezing – he used to sleep naked in the open air.

He used to utter such lovely sentences as, 'Gentler than flowers when it's a question of goodness. Stronger than thunder when it's a matter of principle.' When I expressed my wonder at the harmonious expression of his features, though in his states of trance he must have tremendous experiences, he replied, 'In shallow people the fishes of little thoughts excite great movement, in universal minds even the whale of knowledge stirs up only shallow waves.'

In his neighbourhood, no matter whether he was in trance or in conversation with me, I felt a floating happiness, similar perhaps to the feeling of solemnity in a cathedral. Once I asked him to take me as a pupil. He answered politely, 'It is too soon yet. When you next come.'

Once we came to speak about the often heard legend that Jesus had passed part of his youth in a Buddhist monastery. He is said when fourteen years old to have come to Hemis in Ladakh and not to have returned to the west until he was twenty-eight. What did my holy man think?

'We know and worship Jesus. He is one of the great Bodhisattvas. For him time and space have no significance.' He refused to go into the question more nearly.

The purely external similarity between the ritual and ceremonies of the Catholic and Lamaist religions is surprising – incense, bell-ringing, singing. Many students have attempted to prove an early association between the two religions – perhaps through Thomism. Probably when people want to get close to God there is only a limited range of material means to call on, so that they will necessarily be similar in different religions.

Agricultural terracing, as here at Suikhet, not far from Pokhara, is highly developed among the Nepalese peasants.

172

Khumjung – one of the largest Sherpa settlements. Nearby the Japanese have built a modern hotel with a view of Mount Everest.

Children from Patan. Tiny tots, whom we should still treat as babies, are often put in charge of still smaller children. Smacking is unknown.

More surprising are two legends told among the Lepcha, the aboriginal inhabitants of the former kingdom of Sikkim. This is their story of the flood. Long, long ago the land was inundated by a terrible flood. One Himalayan peak after another disappeared under the water and the people drowned. Only one human couple succeeded in escaping to the summit of the Tedong, the 'sublime horn'. But in height the Tedong, at 2660 metres, is not all that sublime. It is a dwarf among the Hima-

layan giants, but during the flood the gods caused it to grow so that the human couple were saved. The legend does not say whether one pair of every kind of animal accompanied them on to the peak. Like the Noah's Ark family the two Lepchas did not feel safe again till a bird flew by with the usual twig in its bill.

More peculiar is the story of the building of the tower at Sikkim. The Lepchas one day became rather above themselves and wanted to fetch down the sky. They resolved to build a tower so high that by standing on it they could reach heaven with a hook and haul it down. As building material they used earthen pots. They laboured in three groups: the potters who made and fired the pots; the carriers who brought them from the potteries to the building site; finally, the builders who piled the pots on top of one another and built the pile. The work went quickly forward and soon the uppermost builders were so close to heaven that they thought they could get a hook into it. They called down for a hook, a bent rod would do, they said, to pull down heaven. The people on earth thought the work was at an end and that all they now had to do was smash the pots. The tower collapsed and many people were crushed in its ruins. Those who survived could no longer understand one another or speak a common language. In the place where the tower is said to have been built there are now rice fields and again and again new sherds are found there, as evidence of the truth of the tale.

In the eyes of pious Hindus I may have been a deserving pilgrim, because I have visited so many of their sanctuaries. I have made the circuit of the holiest mountain in the world, the Kailas in Tibet, and taken a sin-cleansing – and bitterly cold – bath in the world's most harmonious lake, Manasarowar. Of course I have often been in Benares. In Badrinath, Kedarnath, Yoshimat, and other Himalayan places of pilgrimage, I was often for weeks on end a familiar sight in the temples and to the pilgrims perhaps rather an unwelcome one.

Some years ago my pilgrimages procured me a surprising experience. The aged President of India, Radha Krishnan, was in Vienna on a state visit, and the Indian Embassy gave a reception in his honour. The old gentleman with the face of an ascetic was sitting on a sofa, visibly exhausted, and one after another of the chosen were presented to him for a few private words. When it was my turn I could think of nothing more sensible to say than, 'Your excellency, from your point of view I must be a real holy man, for I have been in such and such and such places.'

Radha Krishnan replied, 'Why, then we're colleagues. Sit down beside me and act as if you were telling me something. I shall nod. I need to rest a little.' I moved my lips and the President nodded from time to time. After some minutes he said, 'I think we mustn't do it any longer. Many thanks.'

I had the longest 'conversation' with him and was correspondingly envied, for Radha Krishnan was rightly considered a great statesman, scholar, and holy man.

There is scarcely a book about Nepal that does not quote Rudyard Kipling's famous couplet,

And the wildest dreams of Kew
Are the facts of Kathmandu.

But Kipling was never in Nepal. No doubt you can have wild dreams in Kew. But just as surely even the wildest fancies will be surpassed in a kind and surprising way by Kathmandu and Nepal.

The area of Nepal is 144,000 square kilometres and it has 13,000,000 inhabitants, half of whom are under twenty years old. It is 800 kilometres long and in its narrowest place from north to south only 144 kilometres wide. On these less than 150 kilometres all the earth's climatic zones meet. In the Terai (wetland) tropical jungle with temperatures of 104°F, on the eight-thousander peaks arctic conditions with 40°F of frost. In between are orange groves recalling Spain, high summer pastures as in Switzerland with shaggy yaks instead of cows, peaks beside which the Eiger is an alluring hill. Imagine, in two hours a bird can fly over every climatic possibility of the earth – in scientific terms the seven geomorphological zones.

With its 2500 kilometres the Himalayan range is twice as long as the Alps, and Everest is twice as high as the Matterhorn. Of the fourteen mountains higher than 8000 metres ten are in the Himalayas, four in the neighbouring Karakoram range.

Page 177
A sadhu, a holy man from Nepal. The Himalayas with their holy sites are a 'social safety-valve' for India. Thousands of pilgrims, with a proportion too of charlatans and people touched in their wits, seek the mountains every year.

Page 178/179
The gigantic figure of the sleeping Vishnu in Budhanilkantha near Kathmandu rests on a bed of snakes in the middle of an artificial pool. It belongs to the early period of Nepalese art.

Page 180
Ladakhi women at the Hemis Feast. The women of Ladakh are very independent and take on all comers in the retail trade. Occasionally, too, for economic reasons they practise polyandry.

Page 181
Braga in the Northern Himalayas. The meagre ploughland is so precious that the houses climb the slopes and are built stepwise above one another.

Page 182 (top and bottom)
The town of Manang in Northern Nepal. The flat roofs are characteristic in the arid climate north of the main Himalayan ridge. The prayer flags are indicative of Lamaism.

Page 183
Reflection of the holy Mount Machapuchare.

Page 184
Small boy from Braga. Hygiene here is not very widespread. All the same the children make a healthy and cheerful impression.

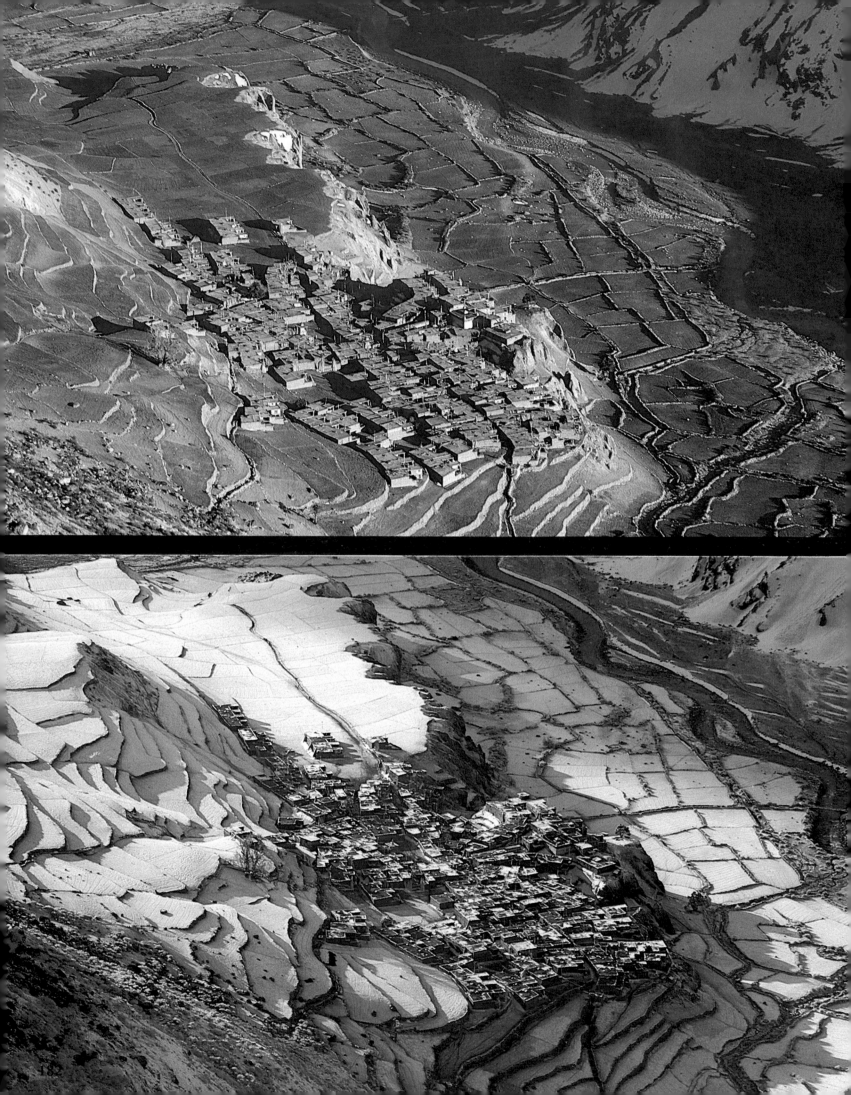

Geologically speaking, the Himalayas are a young 'crushable zone'. The Indian subcontinent floated slowly from the sea where Madagascar now lies towards the Asiatic block. The collision produced folds – the Himalayas. The pressure still continues, and the Himalayas are still growing – the estimates vary between two and sixteen centimetres a year.

Despite their tremendous bulk the Himalayas are not a watershed, as are the Alps in Europe. In terrifyingly narrow gorges some few rivers pierce their main ridge. Dhaulagiri (8167 metres) and Annapurna (8091 metres) are only thirty-five kilometres apart, between them the valley floor of the Kali Gandaki just reaches 2300 metres – an altitude difference of 5800 metres between valley and summit!

The climbers, they say, discovered Nepal, the hippies Kathmandu. The climbers have done a thorough job. The 'decade of the eight-thousanders' began in 1950. In this space of time all the eight-thousanders were climbed, first Annapurna by the French and in 1960 Dhaulagiri by the Swiss. Only the 8046 metre Shisha Pangma limped four years after them, because it lies in sealed off Tibet and the Chinese took their time.

In the climbing of the eight-thousanders a hitherto unknown people became world-famous – the Sherpas. The heroic age of Himalayan mountaineering is over, but the waiting lists for the higher summits become longer and longer, and the Nepalese government is kept busy collecting climbing tolls.

In Nepal live about two dozen different tribes. The Tibeto-Burmese came from the north, the Newar, Thamang, Gurung, Limbu, Rai, Magar, Sherpa, Bhotia. The Indo-Aryans who have the political supremacy today came from the south, as refugees after the conquest of India by Islam in the twelfth century.

Both groups brought their own religion with them, from the north came Buddhism (Lamaism), from the south Hinduism. In the 'Vale of Nepal' the two religions mingle in a cheeringly peaceful manner. Every sanctuary is open to the deities and worshippers of both systems of belief.

Since the people from the north did not want to descend too deeply into the warm valleys, while those from the south had no desire to climb too high, the two groups keep apart from one another, only in the valley of Kathmandu do they mix thoroughly.

First came the Newar and they are there still. They are divinely inspired artists who have turned the valley into a treasure house of breathtaking works of art. They invented the pagodas which were exported by the rulers of those times to China, along with the corresponding architects. The measureless wealth of Nepal derives from the great China-India trade route which passed the ancient royal cities. Occasionally a Chinese emperor would send a trade mission to Nepal to buy jewellery and works of art. The Newar had little share in these riches, they kept their culture intact and were not much concerned who happened to be ruling them at a given time. These were first of all the Kirate, who had at least had the advantage that they came from the north and were in a manner of speaking cousins. It was in their time that Buddha was born in Lumbini (what is today Nepal), and a little later the free-thinking emperor Ashoka visited the Vale. He brought Buddhism and Indian influences. Then came the Lishawi, the Thakur, the Malla, all from the south. They promoted the arts, fought one another, concluded political marriages with Peking and Lhasa, trade flourished, and the Newar built, forged, and carved.

The more recent history of Nepal reads like a mixture of Lehar operettas and Shakespearian history plays A succession of intrigues and murders, both with childlike charm and frightful cruelty. The Malla were replaced by the Gurkhas and the victorious Prithvi Narayan did not forget that the village of Kirtipur had offered him obstinate resistance. After his victory he had most of the men killed and the noses of the remainder cut off. Since then Kirtipur has been called the 'village of the noseless men'.

The victorious royal house in its turn succumbed to intrigue and was replaced by the Rana dynasty (1846–1951), which made the office of prime minister hereditary and shut up the king, as a reincarnation of the god

Vishnu, in a golden cage. Occasionally the kings were allowed to appear at religious feasts or read a proclamation, but otherwise they were powerless.

In 1950 there began under partly obscure, partly romantic circumstances, strongly supported by India, a socialist revolution, which put an end to the 'Ranarchy' and set the king back on his throne. A German hospital nurse (shades of *The King and I*!) conducted the bewildered god-king, despite Rana surveillance, to the Indian embassy. From there he was flown to India, and returned after a period of confusion as rightful king. Today his grandson sits on the throne.

Once a year the king has his sovereignty confirmed by the living goddess of the Vale, the Kumari. The Kumari is a little girl chosen from the goldsmiths' caste. She can only remain goddess until she bleeds, whether from a pricked thumb or the onset of puberty. Then another goddess is chosen. The lot of the child goddesses, despite the pomp with which they are surrounded, is not all that happy, because it is supposed to be unlucky to marry a former Kumari.

Of course there are modern things too, UN, UNESCO, 'development aid', in the Vale, but they have a job to establish themselves in a 'land between heaven and earth'.

Recently an Alpine periodical organised a poll on the question 'Which of the world's mountain ranges do you like best?' I answered without a second thought, 'The Himalayas.' When the reporters asked me my reasons, the answer again seemed to come of its own accord, 'I like the Himalayas because they're religious mountains.' I could almost have said Christian mountains.

I knew I had found the right explanation. Of course I have talked about the scenery – the rhododendron forests in Nepal and the fields of edelweiss by Nanga Parbat – or about the hospitality of the mountain people, but these would only have been aspects which one can find in other mountain ranges, though perhaps now diluted. But the unique thing about the Himalayas is their godlike quality.

Of course one can argue that mountains are rock and ice formations, which can at most be majestic, repellent, or terrifying, but not god-like. That is to ignore the feelings of the people in their neighbourhood. The religiosity of the Asian peoples, especially the Hindus, is deep-rooted and today still rules their daily lives. Since they mostly live in tropically hot, often arid lands, glacier mountains must to them seem like forms from another world, like heaven itself, in which their gods do as they please. The great rivers emerging from the Himalayas water the north of India. Their water ensures the harvest, and every harvest engenders new life. Life is a gift of the Himalayas or of the gods enthroned there. No wonder then that hundreds of millions of Indians think of these mountains with worship and love.

Let me return to the objection that mountains are material forms of the earth's surface. Today we know that our thoughts and wishes are currents and vibrations which can be recorded and measured. If these forces are concentrated a millionfold on one area or one point they will engender there a special aura. And perhaps it is this aura which registers as the magic of the Himalayas.

If today tens of thousands of tourists travel every year to Nepal and Lhadak, it is not only to wonder at mountains, buildings, and works of art, but to participate in that simple life which still produces contented faces. Perhaps the fascination which the Himalayan lands exert on us is in fact our own search for a Shangri La, a happiness without wishes, and the longing to lead an ageless life.

The unclimbed holy mountain of Jobo Lhaptshan, 6440 metres high. Many mountains are considered to be the abodes of particular gods. Climbers must undertake to turn back a few metres below the summit, so as not to disturb their quiet. There are many peaks for which no permits are issued.

Every mountain-climber who attempts an especially striking peak in the Himalayas is there looking in his own way for his Shangri La. Sepp Jöchler, who together with Pasang and me climbed Cho Oyu, said, 'After an eight-thousander one is no longer the same person.' I think the experience of a high summit cannot be put more tellingly or more succinctly. You have found your Shangri La.

The pilgrims on the pilgrims' way to Badrinath, where the holy Ganges has its source, are mostly poor peasants from one of the countless villages of the sweltering Indian plains. They know how to be miserly with their scanty means, for temple guardians, dealers and mountebanks try by all the tricks of their art to fleece them thoroughly. They find themselves in a world which not only makes them happy but also fills them with terror. They are careful and timorous.

And yet I shall never forget the look of an unimaginable happiness on the face of one old pilgrim. It was early morning and he was squatting in the door of his modest lodging and gazing at the summit of Nilkantha, the 'Blue-throat', which was beginning to gleam in the first light of day. I do not know what god lives in this mountain or which one the peasant was adoring, at any rate his face glowed with a happiness which most of us can only hope to experience.

Inside these prayer wheels are prayers printed on strips of paper.

The Mahakala, tutelary deity of the Buddhist faith, here carved in stone, is a manifestation of the goddess Durga and belongs to the group of eight mother-goddesses.

196

Herbert Tichy
was born on June 1st, 1912 in Vienna. His father was a lawyer. He studied geology at the universities of Vienna, Lahore, and Benares. His doctoral thesis in Vienna in 1937 was on geological problems in the Himalayas. In 1938 he worked as geologist and journalist in Alaska, later as an oil geologist in Austria, Germany, and Poland.

His travels started in 1933 when he rode his motor-cycle from Vienna to Bombay, and he has since travelled extensively all over the Far East, as well as in Africa. Altogether he has visited the Himalayas and the Hindu Kush seven times and became particularly known for his small-scale expeditions, when he wandered for months on end through the mountains, often only accompanied by one local porter. The climbing of Cho Oyu (8153 metres) with the smallest party at that time (1954) to venture such an undertaking, three Austrians and ten Sherpas, was a triumph acknowledged all over the world. Tichy continues to this day as a writer, reporter and explorer in Africa and Asia, and has received numerous distinctions for his scientific, climbing and literary achievements. His books translated into English are *Cho Oyu: by favour of the gods* (1957) and *Himalaya* (1971).

*Translated from the German
by Christopher Holme*

198

on the Trail
of Jacques Cousteau
or Paradise
under the Ocean

Leni Riefenstahl

Leni Riefenstahl

These days people often ask me what the sea means to me. I would prefer to replace the word 'sea' by 'water', which is to me a very familiar element. Why? That goes back far into my past. My parents had a building-plot outside Berlin, and there I was thrown into the water at the tender age of five. We had a boat, and I swam as often as I had a chance. At ten years of age I joined a swimming club called 'Nixe' and took part in a few competitions. We also developed a special little sport – diving for plates and retrieving them. And so I early became very familiar with water.

Later I became a real long-distance swimmer, and I also learnt sailing, before the mountains entered my life. Decades followed, during which I lived in and with mountains. Mountain films such as *The Holy Mountain* (1926), and *The White Hell of Piz Palü* (1927), claimed my total attention, and then there was *Storm over Mont Blanc* (1930), *The White Inebriation* (1931) and *The Blue Light* (1932). Finally came the film on the Greenland expedition *SOS Iceberg*. As time passed I completely lost touch with the water.

It was almost half a century, before I returned to the water, this time in fact, to the sea. I was in Africa and went to the Indian Ocean to recuperate after some exhausting photographic work. Together with a group I tried swimming with a snorkel. It was ebb, not very deep water; only an inch or so beneath me a whole new world began.

To me this was a fated turning-point. I was amazed by what I saw: by the fish, by the corals, by the colourful palette extending in front of my eyes. I was like a child enjoying Christmas – I wanted to experience it all over again. But at that moment I had to leave. One year later I returned to the Indian Ocean. I had a secret wish to see this world of reefs in iridescent colours, not only from above, but also from below – like a fish among fish, as it were.

But for that I would have to learn to use diving equipment, to breathe through it, and to move with it. I had to take an examination. But how can one pass a diving test together with

nothing but youthful twenty and thirty-year-old athletes, when one is seventy-one years of age?

I resorted to a simple trick: made myself twenty years younger and used my other name, the name in my passport: Mrs. Jacob. I passed the test and since then I have been lost to diving.

In those days I was not yet thinking of taking photographs – I just wanted the experience: to dive and be able to watch this far-stretching unrevealed world, gliding effortlessly down and up through the water. I have practised many sports in my life, I have climbed mountains and have enjoyed great happiness doing that. But diving surpassed it all.

To be able to explain why, I shall have to recall the state I experience every time I glide into the water from the boat. I can hardly wait for this moment! I'm enveloped by a world which has no relation to the earth 'up there'. A new world. I'm floating, through this blue-green world. At first I don't recognise a thing apart from the feeling of total lightness. Then, slowly, I start to orientate myself, my eyes discover contours, a reef appears in the distance, the first coral stocks move gently in the stream.

With a few strokes of the fins I drift towards the corals. Now I recognise the multitude of life in there, the home of innumerable reef fishes, dancing in shoals. I could stay for hours in front of these coral settlements, just watching. I always feel sad when the pneumatic air gauge indicates that the air in the diving equipment is running out and I must get up again.

I became aware of a wish to photograph what I saw. When I went to dive in Port Sudan on the Red Sea, just after completing a land expedition, I brought a camera along, for the first time. The camera was uncomplicated: it was an automatic, housed in a plastic case. The pictures were certainly, as could have been expected, of average quality.

Page 201
Leni Riefenstahl, taking pictures at a depth of thirty metres of the shark cage by the Shab Roumi reef. This cage was used by Cousteau and his divers fifteen years earlier for some of his shark pictures. In the passage of time the cage has been overgrown by colourful corals and sponges.
Sudan – the Red Sea.

Pages 202/203
This picture shows clearly that the coral only reveals its splendid colours when hit by an underwater light. Without this, the diver sees the submarine landscape in blue-green colours.
Shab Roumi Reef – Sudan – the Red Sea.

Pages 204/205
White-fin lionfish PTEROIS RADIATA *(20 cms)*
Shab Roumi Reef – Sudan – the Red Sea.

Page 206 top
Sweetlips fish PLECTORHYNCHUS ALBOVITTATUS *(45 cms)*
Sanganeb – Sudan – the Red Sea.

Page 206 bottom
Anemone fish AMPHIRION BICINCTUS *(8 cms) lives in symbiosis with a poisonous anemone.*
Shab Roumi Reef – Sudan – the Red Sea.

Page 207
Squirrel fish HOLOCENTRUS SPINIFER *(40 cms). The small blue fish are cleaner fish.*
Shab Roumi Reef – Sudan – the Red Sea.

208

Leni Riefenstahl in the tracks of Jacques Cousteau, diving with the aqualung developed by Cousteau in the Red Sea near Port Sudan. In the background Cousteau's 'garage' for submarine vehicles.

Pages 208/209
Barracuda SPHYRAENA BARRACUDA (1 m) under a shoal of herrings HARENGULA CLUPEOLA (10 cms). North Eleuthera – the Bahamas – the Caribbean.

Page 210
Soft coral DENDRONEPHTHYA Sanganeb – Sudan – the Red Sea.

Page 211 top
Squirrel fish HOLOCENTRUS SPINIFER (40 cms)
Shab Roumi Reef – Sudan – the Red Sea.

Page 211 bottom
Soft coral DENDRONEPHTHYA with sponges
Shab Roumi Reef – Sudan – the Red Sea.

Pages 212/213
These flowers are polyps of the coral TUBASTREA, with which it takes in plankton nutrition. This is done at night.
Sanganeb – Sudan – the Red Sea.

Pages 214/215
Coral trout CEPHALOPHOLIS MINIATUS (50 cms)
Turtle Bay – Kenya – Indian Ocean.

Page 216
Horn coral LOPHOGORGIA with extended polyps. The little fish are lyre-tailed coral fish ANTHIAS SQAMIPINNIS
Sanganeb – Sudan – the Red Sea.

I realised that I would not be able to reproduce my own fascination with that camera. I had to start afresh. First I bought books written by leading submarine photographers, and tried – to begin with at least in theory – to penetrate these quite different conditions. But theory remains grey and only assumes colour in the literal sense of the word when, finding yourself eye to eye with an emperor fish or a large, glistening coral trout, you can apply the theory in practice, that is, in a picture.

I managed to do this during a long stay in the Caribbean, after diving in the Indian Ocean and the Red Sea. I now used a Nikonos, which is the most popular submarine camera amongst photographing divers and diving photographers. It needs no protective case, can be used both in water and on land and is in many instances indispensable.

Calypso, *Cousteau's research ship, originally a minesweeper, built by the British in America in 1942. She is 43 m long, 7 m wide and 329 tons, with room for around thirty people and the very bulky scientific equipment.*
The maiden voyage was in 1953 in the Red Sea.
In front of the bow is the famous 'false projection', a metal shaft containing a watch room with five portholes where observations can be made and filmed underwater without effort, even while the ship travels.
Calypso has travelled hundreds of thousands of nautical miles on her cruises over all the seven seas. One result of these voyages are the Cousteau television films, which have been

218 *transmitted all over the world.*

A diver with an aqualung and a helmet telephone. Today's free diving equipment has revolutionised both scientific and sporting diving. The breathing equipment developed by Cousteau and Gagnan lets the consumed air into the water at once and adjusts the air flow to the breathing rhythm of the diver. The containers of strongly compressed pneumatic air are carried by the diver on his back. The fully automatic equipment is as simple as it is safe to use.

I noticed, however, that the pictures improved if I used a mirror reflex camera for the pictures as I imagined them. I bought the Nikon model in a submarine case. It soon became clear to me that I really ought to work with different lenses, according to each motif that offered itself to me. Apart from the mirror reflex camera, I used my two Nikonos cameras, which I attached to a line before diving, so that I could always find them when I needed them.

I naturally also realised the decisive importance of the flashlight. Only the artificial light, brought by the diver, will release the incredible multitude of colours, which are hidden by something like a blue filter to eyes not normally designed for seeing underwater. My first pictures taken with flashlight were still too shallow, but then I learnt to apply the flash or

flashes from the side or above, in order to create depth and release the objects from the background.

Finally my eyes got used to imagining the colours in the blue-green water. I now knew where the gorgeous colour range of sea slugs, sponges or soft corals was waiting for me. I went looking for motifs with a little hand spotlight and awakened the hidden world of colours resting there.

What I find most interesting in the sea is not so much the big things, the so called sensations, but rather the little things, the micro-worlds, which contain such an infinite number of other things.

Occasionally it is also the landscape that attracts me: raising my eyes from a submarine cave and staring out over coral fields, fire and brain corals into the open, blue land without horizons. Then it is naturally important that you

release the foreground plastically with your flashlight – this is one of the secrets of good submarine photography. You cannot wait for a natural light change, as you do in land photography.

But the main interest is the still life quality of the coral gardens, the picturesque little worlds, one could say, underneath the sea surface. It is also as if the world 'blossomed' down there. This has certainly nothing to do with the seasonal blossoming on land. After having seen a lot of things one learns that all these apparently blossoming structures really are the functional organs of different animals. The 'blossoming sea' created by tube worms is their feathery cups stretching out for nutrition.

In the same way the polyps of the leather coral 'blossom' when they open their tentacles, like the various soft corals. This happens mainly at night, the main time of activity for these organisms. That is when a photographer is most challenged. One's nerves couldn't be more tense!

As a diver you have to find your way about in a completely unfamiliar environment, in which normal dimensions no more apply. In the darkness sleeping fish or quite lively coral communities emerge. What motifs! With all this it's easy to forget that they are not without danger to man. But for some reason, I always feel safe in water.

Still, even with my best pictures I have a feeling that this is but a beginning. There is so much to learn if you want to exhaust the possibilities of submarine photography. In this seeming dream world, a person without religious faith could well find it underwater, because creation here shows

The diving saucer, developed by Jacques Cousteau in several models. The SP 350 for two people is equipped with a film camera, a photo camera, a hydraulically powered seizing device and a collecting basket. It has made hundreds of successful dives. The 'Seefloh' (SP 1000) for one person carries two external remote control cameras and recording equipment to record submarine noises. Other models can dive at a depth of 1200 metres and carry three people. The speed is on average two knots.

you things that cannot be rationally understood.

As I said, as far as I am concerned, there is a whole unexplored world of new motifs and new possibilities of interpretation underwater.

But I would like to add one thing. In 1963, at the time when I travelled alone through Africa, including the Sudan, a magnificent experiment took place actually in the Sudan, twenty-five kilometres north of Port Sudan. This submarine experiment now forms part of the history of submarine exploration. It was at this time that Cousteau started his enterprise with a submarine 'village' at the coral reef Shab Roumi.

Fifteen years later I was lucky enough to dive there. Even today the large shark cage, overgrown with corals, remains, and the submarine house, now a favourite motif for divers who go there. I found it all fascinating, above all, it was interesting to see how soon nature had overgrown and changed what man had built. Today the steel constructions are covered by coral gardens.

Cousteau's submarine 'village' at Port Sudan. In the submarine garage on the right, a diving saucer is just floating in (middle foreground). In the background is the housing for seven divers, in which they spent a month at a time. This experiment was one of a series seeking to establish the possibilities of living underwater (Precontinent I, II and III).
In 1965, six divers spent three weeks at a depth of 100 metres at Cap Ferrat on the French Mediterranean coast.

222

Jacques Cousteau and his men onboard Calypso. *A captain with experience of all seven seas and a great researcher, Cousteau knows how to guide his assorted team of assistants and crew members successfully on each new cruise.*

When you have experienced this vitality of the world of the sea, you can only hope with all your heart that it will be preserved for all of us and for future generations. The submarine world has grown so slowly, but now materialism, brutality and careless-ness can destroy or seriously threaten to destroy it in less than one gener-ation's time.

Since only 1973, the years I have been diving, I have already had to wit-ness the destruction of the reefs.

No one should take anything away from the water if he doesn't need it. Corals and mussels should be left where they have grown. And fish should not be harpooned by divers.

Even so this is, sadly, just the small destruction. The great destruction we are causing is an entirely different matter. I am a whole-hearted sup-porter of the battle fought by Cousteau today for the preservation of the 'blue planet'. I would like to ask all readers to support this battle, wher-ever it is fought.

What has grown up over centuries can so quickly be destroyed. After all, the conditions under water, for ex-ample where corals and fishes and sea shells are concerned, are so unique in their function, in life as a whole, and no such density really exists on land. Conditions there are still close to the primordial state.

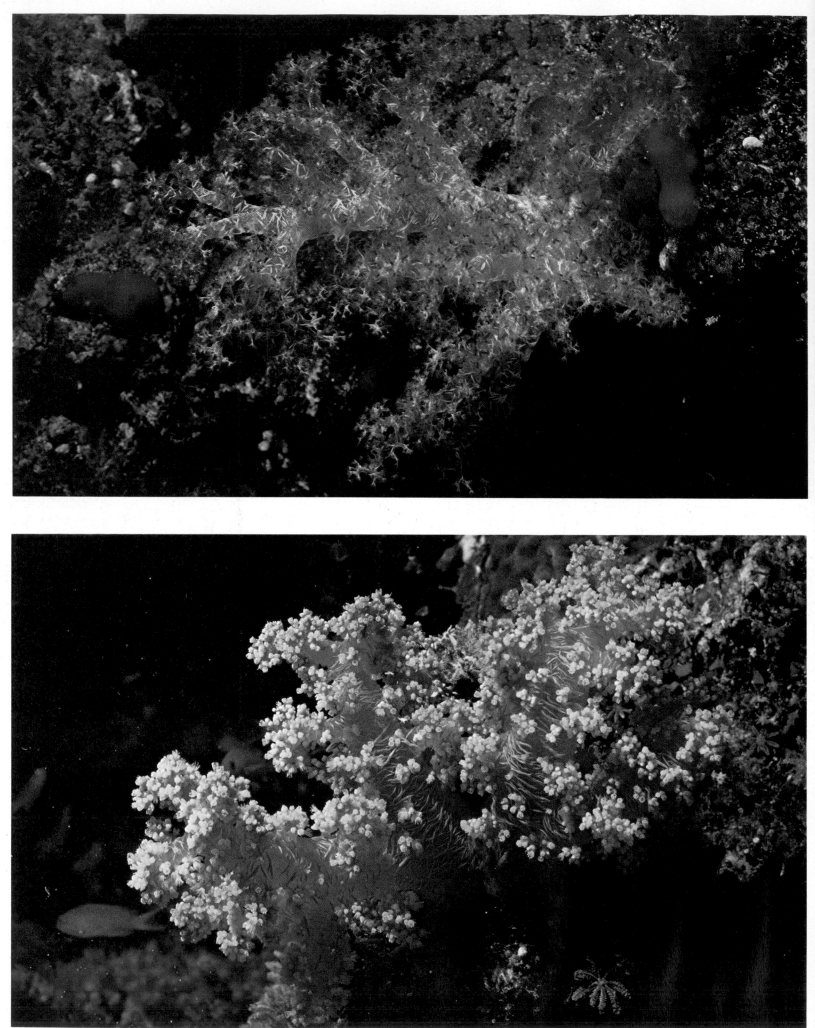

232

Jacques-Yves Cousteau
was born in *Saint-André-de-Cubzac* (Gironde) in France on June 11th, 1910, the son of a solicitor. He chose the career of a naval officer and was in the French Navy from 1930 to 1956, when he retired, a lieutenant commander.

His lifelong interest in swimming, diving and exploring the submarine world started as early as 1936. In 1947 he beat the world record in free diving (without diving equipment) at a depth of 91.5 metres. In 1946, the first diving equipment using pneumatic air (the aqualung) was designed by him and produced for the first time. After that, Cousteau developed several other pioneering diving devices, diving vehicles and three submarine laboratories.

Cousteau became world famous through his expeditions on the research ship, *Calypso*, which is also a floating laboratory. He spends an average four months a year at sea, and his numerous television series such as *The Silent World*, are highly loved and acclaimed in many countries. He has dived under icebergs, he has filmed sharks, and has thoroughly investigated all the seven seas.

In 1970, after taking *Calypso* for a three and a half year voyage, travelling more than 250,000 kilometres through the Mediterranean, the Indian, Atlantic and Pacific Oceans and the Caribbean, Cousteau declared that he considered the fauna and flora of the sea very seriously threatened by pollution, and that forty per cent of submarine life had been destroyed within the past twenty years.

Cousteau had earlier on participated in raising ancient treasure and ships from the Mediterranean off the African coast. He also searched for the legendary Atlantis – which is said to have sunk into the sea many thousand years ago.

*Translated from the German
by Ann Henning*